Structure & Statistics
In Crystallography

Structure & Statistics In Crystallography

Proceedings of the Symposium on Crystallographic Statistics held in Hamburg, West Germany in August 1984 in the course of the Thirteenth International Congress of the International Union of Crystallography.

Edited by

A.J.C. Wilson F.R.S.
University of Cambridge, England

Adenine Press, P.O. Box 355
Guilderland, NY 12084 USA

Adenine Press
Post Office Box 355/340
Guilderland, New York 12084

Copyright ©1985 Adenine Press Inc.

Cover illustration: Trial and actual structures of diphenyl sulphoxide obtained by refinement; From: Milledge et al., this volume.

Library of Congress Cataloging-in-Publication Data
Symposium on Crystallographic Statistics (1984:
  Hamburg, Germany)
  Structure & statistics in crystallography.

  Bibliography: p.
  Includes index.
  1. Crystals—Congresses. 2. Crystallography—
Statistical methods—Congresses. I. Wilson, A.J.C.
(Arthur James Cochran), 1914- . II. International
Union of Crystallography. International Congress (13th
: 1984: Hamburg, Germany) III. Title. IV. Title.
Structure and statistics in crystallography.
QD921.S937   1984   548'.7   85-26680
ISBN 0-940030-10-1

All Rights reserved. No part of this book may be reproduced, stored in a retrieval system or transmitted in any form or by any means: electronic, electrostatic, magnetic tape, mechanical, photocopying, phonographic recording or otherwise, without permission from the publisher.

*Set in type in the United States by Word Management Corporation.
Printed in the United States by Hamilton Printing Company.*

CONTENTS

Introduction
    A.J.C. Wilson   v

Intensity Statistics

Crystallographic Statistics and the Development of Direct Methods
    J. Karle   1

Fourier Series and Other Representations of Crystallographic Probability Density Functions
    G.H. Weiss, U. Shmueli, J.E. Kiefer and A.J.C. Wilson   23

Effect of Heavy Atoms on Intensity Statistics of X-ray Diffraction from Crystals
    D. Pradhan, S. Ghosh and G.D. Nigam   43

Centric, Bicentric, and Partially Bicentric Intensity Statistics
    U. Shmueli and G.H. Weiss   53

Non-Crystallographic Translational Symmetry: Effects on Diffraction-Intensity Statistics
    G. Cascarano, G. Giacovazzo and M. Luić   67

A Best Test to Distinguish between X-ray Intensity Distributions for Space-Group Determination
    S. Parthasarathy and N. Elango   79

Effect of Statistical Fluctuations and Systematic Errors on Intensity Distributions—Which Way Forward?
    A.J.C. Wilson   87

Estimation of Parameters

Precision and Accuracy in Structure Refinement by the Rietveld Method
    E. Prince   95

Entropy Maximization: An Alternative to Squares Minimization
    D.M. Collins   109

Information and Crystal-Structure Estimation
    S.W. Wilkins, S. Steenstrup and J.N. Varghese   113

A Method for the Systematic Modification of Least-Squares Weights to Account for Residual Error
    Wang Hong and B.E. Robertson   125

The Variances and Covariances of Measured Intensities in Precise Lattice-Constant Determination by the Bond Method
    E. Gałdecka   137

Precise and Accurate Estimation of Crystallographic Parameters by Maximum Likelihood and Min-Max Methods
    G.B. Mitra, Rabea Ahmed and Prabal Das Gupta   151

The Influence of Individual Reflections on the Precision of Parameter
Estimates in Least-Squares Refinement
   E. Prince and **W.L. Nicholson** 183

Principles Involved in the Development of Expert Systems for
Data Acquisition
   H.J. Milledge, M.J. Mendelssohn, C.M. O'Brien and **G.I. Webb** 197

Index 223

Introduction

The greater part of this volume consists of the papers given at a symposium on Crystallographic Statistics held on 17 August 1984, in the course of the Thirteenth International Congress of the International Union of Crystallography. Dr Jerome Karle, President of the International Union of Crystallography (1), was invited to review 'Crystallographic Statistics and the Development of Direct Methods', and an extended version of his symposium talk is the first paper in the book. Other authors, acknowledged experts in their fields, were invited to review specific topics within the area of crystallographic statistics, and the rest of the time allotted to the symposium was filled with closely related contributed papers. In addition, a few papers that could not be accommodated within the symposium but were presented otherwise within the Congress have been included. This volume may thus claim to represent the 'state of the art' of crystallographic statistics in mid-1984. The papers fall into two groups: those dealing with intensity distributions, and those dealing with estimation of parameters.

A similar symposium (2) was held at a Congress of Crystallography in 1981, and in commenting on it I drew attention to three problems then outstanding:

(i) bias in the estimation of parameters;

(ii) the effect of correlations of atomic positions on the expressions for the probability distribution of intensities; and

(iii) the functional form of the distribution of really large intensities.

In the intervening years the third of these problems has been solved for the space group $P\bar{1}$ with some progress for other space groups (3,4); the solution is reviewed briefly in the paper by Dr G.H. Weiss and his co-authors (5). The main part of Dr Weiss' paper, however, deals with the representation of probability distributions by Fourier series (4,6); these are more convenient than the Gram-Charlier or Edgeworth series used in most earlier work, and are for practical purposes 'exact' representations of the desired functions. Several contributed papers on probability distributions follow that of Dr Weiss. Shmueli & Weiss (7) and Cascarano, Giacovazzo & Luić (8) discuss the effects of additional non-crystallographic symmetry, and Parthasarathy and Elango (9) propose a likelihood-ratio test for distinguishing between distribution functions. Pradhan, Ghosh & Nigam (pp. 43-51 below) discuss the effect of heavy atoms.

There is no substantial progress to report on the second of the above problems. Estimation of parameters, however, forms the topic of the second section of this volume, as it did of the symposium. Dr E. Prince reviews 'Precision and Accuracy in Structure Refinement by the Rietveld Method' (10), a matter about which there has been some controversy in recent years. This paper gave rise to a lengthy oral discussion after the official end of the symposium; the discussion (11), if not clearing up the matter to the satisfaction of all participants, seemed to show that the disagreements were largely a matter of terminology rather than statistics or science. The title of Dr Prince's paper was dictated by the controversy, but most of his considerations apply to other methods of structure refinement as well. His paper is followed by several others on aspects of parameter determination, with occasional sideways glances at bias. New methods of estimating parameters, based on entropy maximization, are discussed by Collins (12) and Wilkins, Steenstrup & Varghese (13), while Wang Hong & Robertson (14) deal with procedures for optimizing weights to deduce the estimated standard deviations (with somewhat more than a sideways glance at bias). Two papers with a practical slant are Gałdecka (15) on lattice-parameter determination by the Bond method and Mitra, Ahmed & Das Gupta (16) on the lattice and other parameters of a variety of substances. The final papers, by Prince & Nicholson (17) and Milledge, Mendelssohn, O'Brien & Webb (18), are concerned with optimizing the use of available time for increasing precision of the estimates of chosen parameters.

Many problems remain outstanding, and perhaps the important ones are those of which we are not yet conscious. However, the following seem obvious:

(i) The effect of correlation of atomic positions on the probability distributions, already mentioned, has been investigated only as far as the second moment (19).

(ii) Fourier representations of probability distributions are available only for the lower-symmetry space groups (all triclinic and monoclinic, most orthorhombic, a few others). There is no fundamental reason why the remaining space groups cannot be treated, but the mathematics becomes tedious.

(iii) In spite of the many papers (20) in which the subject is mentioned, avoidance of bias in the estimation of parameters is not fully understood. There are two types of bias affecting the refinement process, one resulting from imperfections in the mathematical process as such, and the other resulting from defects in the model. 'Defects in the model', or some of them, may alternatively be regarded as uncorrected systematic errors.

(iv) Observed and calculated values of physical quantities—in the present context probability distributions or structure factors—always differ from one another to a greater or lesser extent. Tests are needed to decide whether such differences can reasonably be attributed to statistical fluctuations or sampling effects, or whether they indicate the presence of systematic errors (or defects in the model) that are comparable with, or larger than, the statistical fluctuations (sampling effects). There are many papers (21,22,23) on this subject, but no method has become routine.

Perhaps progress in these matters can be reported by the time of the next Congress of Crystallography, scheduled for Perth (Australia) in August 1987.

I wish to express my thanks to Professor Dr H. Burzlaff, the co-chairman of the symposium, and to Professor Dr U. Bonse, the Chairman of the Programme Committee of the Congress. Both had to cope with the sometimes difficult requirements of the symposium participants and Chairman.

A.J.C. Wilson

Notes and References

1. Dr Karle's term of office expired on 18 August 1984; the President is now Professor Dr Theo Hahn.
2. S. Ramaseshan, M.F. Richardson and A.J.C. Wilson (eds.), *Crystallographic Statistics: Progress and Problems*. Bangalore: Indian Academy of Sciences (1982).
3. A.J.C. Wilson, *Acta Crystallogr.* A *39*, 26-28 (1983).
4. G.H. Weiss and J.E. Kiefer, *J. Phys.* A *16* 489-495 (1983).
5. This volume, pp. 23-42.
6. U. Shmueli, G.H. Weiss, J.E. Kiefer and A.J.C. Wilson, *Acta Crystallogr.* A, *40*, 651-660 (1984).
7. This volume, pp. 53-66.
8. This volume, pp. 67-77.
9. This volume, pp. 79-85.
10. This volume, pp. 95-103.
11. This volume, pp. 103-107.
12. This volume, pp. 109-112.
13. This volume, pp. 113-124.
14. This volume, pp. 125-136.
15. This volume, pp. 137-149.
16. This volume, pp. 151-181; compare ref. (22).
17. This volume, pp. 183-195.
18. This volume, pp. 197-211.
19. A.J.C. Wilson, *Acta Crystallogr.* A *37*, 808-810 (1981).
20. Some references are given by A.J.C. Wilson, *Acta Crystallogr.* A *35*, 122-130 (1979).
21. The earliest directly relevant work seems to be that of K.E. Beu, F.J. Musil and D.R. Whitney, *Acta Crystallogr. 15*, 1292-1301 (1962); for further references see (22). Although Beu *et.al.* used maximum-likelihood methods, and considered lattice parameters only, their arguments are readily adapted to least squares and structural parameters (22).
22. A.J.C. Wilson, *Acta Crystallogr.* A *36*, 937-944 (1980).
23. See (6), particularly Appendix B.

Structure & Statistics in Crystallography, ISBN 0-940030-10-1,
Ed., A. J. C. Wilson F. R. S., Adenine Press, ©Adenine Press 1985.

Crystallographic Statistics and the Development of Direct Methods

J. Karle
Laboratory for the Structure of Matter
Naval Research Laboratory
Washington, D. C. 20375, U.S.A.

Abstract

The foundation mathematics for direct methods was developed from the inequality theory that is associated with the non-negativity of electron density distributions in crystals and its probabilistic implications. The earliest probability considerations that were introduced into crystal structure analysis were the intensity statistics of Wilson. With them, he showed how experimental data could be properly scaled and corrected for positional disorder and how tests could be developed to distinguish the presence or absence of a center of symmetry. He and others also developed insights into how the effects of special positions, a variety of regularities within the unit cell and the presence of heavy atoms can affect the statistical distribution of diffracted intensities. The probability formulas associated with relations from inequality theory that provide the mathematical basis for procedures for phase determination can be considered as an extension of the probability theory associated with intensity statistics. In fact, the application of the joint probability distribution to the development of phase-determining formulas, the first application of probability theory to crystal-structure determination, arose in just that way. The ensuing article is devoted to developing this theme. It discusses why the phase problem is solvable by use of diffraction intensities despite the fact that phases are lost in the recording of the intensities and describes the early work in intensity statistics. This is followed by a historical discussion of the development of phase-determining formulas through the inequality and probability theories, a description in some depth of the probabilistic implications of inequality theory and, finally, some summary thoughts on the subject of crystal-structure determination by means of direct phase determination, the so-called 'direct methods'.

Introduction

Direct methods for crystal structure determination arose from two kinds of theoretical techniques applied to the structure-factor equations, inequality theory and probability theory. In the application of these theories, certain features of crystal structures and diffraction experiments played an important role. These features are the non-negativity of electron density distributions, the discreteness of atoms that can be extended in such a way that the atoms act as if they were essentially point atoms, and the great overdeterminacy of a diffraction experiment in which the number of independent diffraction data far exceeds the number of unknown structural parameters.

The concept of direct methods in crystal-structure determination implies the initiation of a process for determining the phases of the amplitudes of scattering from the measured intensities that does not depend on the introduction of previously known structural information. This concept distinguishes direct methods from Patterson methods since the latter are initially concerned with the gleaning of structural rather than phase information. The presence of heavy atoms in a structure is usually useful because the probability measures associated with phase determination are normally enhanced. However, if the heavy atoms are used in a special way that makes use of structural information for the heavy atoms, for example, the heavy-atom method or the techniques of isomorphous replacement or anomalous dispersion, then these methods are clearly distinguished from those that are called direct methods.

There were early attempts to obtain structural information or phase information from the structure-factor equations. Ott (1) derived relationships among the structure-factors and atomic positions by use of the structure-factor equations and in some simple cases he showed that these relationships could be used to obtain atomic coordinates directly. Banerjee (2) used Ott's results in a trial-and-error self-consistency routine for finding the signs of structure-factors. The number of trials increased rapidly with complexity and errors in the observed data entered in such a way as to again limit applications to the simpler structures. Avrami (3) worked with the equations that relate intensities to interatomic vectors. The solution of these equations was given in terms of the roots of a polynomial equation whose degree increases with great rapidity as the crystal becomes more complex. This fact and accuracy problems associated with the experimental data limited the general applicability of this approach. Little more was done until the development of appropriate inequality and probability theory.

The initial role of probability theory in crystallography was to describe the statistical properties of diffracted intensities and to detect the presence or absence of a center of symmetry. Statistical studies of diffracted intensities were extended a little later to include the effects of symmetry elements, special positions, heavy atoms and unusual structural features such as non-crystallographic symmetry. Probability theory from which phase information may be obtained from knowledge of structure-factor magnitudes was also developed a little later, motivated by the results of inequality theory. The immediate goal of inequality theory, which arises from the non-negativity of the electron density distribution in a crystal, was the development of relations between phases and structure-factor magnitudes. The inherent probabilistic characteristics of the inequality relations facilitated the mathematical realization of the connection between probability theory and inequality theory.

In an ordinary diffraction experiment, the diffraction intensities appear to have lost all phase information. The intensities are proportional to the squares of the magnitudes of the structure-factors only and the individual phase values disappear in the recording of the intensities. Why then should it be expected that phase information is obtainable from the measured intensities? The reason is that despite the loss of phase information in the collection of the diffraction intensities, phase information is still contained in these intensities. Insight into the reason for this derives from a consideration of the structure-factor equations and recognition that the unknown quantities in the equations, the atomic positions and the phases of the structure-factors, are greatly overdetermined by the number of data that can be measured in a diffraction experiment. It was the recognition of this overdeterminacy that strongly motivated the search for practical formulas for phase determination.

Crystallographic Statistics

Statistical concepts were introduced into crystallography by Wilson (4) who demonstrated that intensity data could be corrected for vibrational effects, actually positional disorders of all sorts, and placed on an absolute scale by use of a statistical relationship between the intensities and the atomic scattering factors. This relationship has been used ever since the inception of direct methods to prepare the measured data for further analysis. Wilson pointed out that for intensities that were already on an absolute scale,

$$\langle I_h \rangle = s_2 = \sum_{j=1}^{N} f_{jh}^2 \tag{1}$$

where f_{jh} is the atomic scattering factor for the jth atom in a unit cell containing N atoms. He further noted that the observed intensities I_{hobs} will differ from the I_h by a scale factor k, which is independent of the scattering angle, and by a temperature factor. Wilson proposed a calculation to effect the determination of the scale-factor and the correction for positional disorder. He suggested that the reflections be divided into p groups covering ranges of values for $s = \sin\theta/\lambda$, where θ is the Bragg angle and λ is the wavelength of the experiment. Within each group the ratios

$$\frac{\sum_{sj-1 \leq s \leq sj} s_2(s)}{\sum_{sj-1 \leq s \leq sj} I_{\text{obs.}}(s)}, \quad j=1,2,\ldots,p \tag{2}$$

are computed. If the temperature factor is of the form $\exp(-Bs^2)$, then $I_{hobs} = kI_h\exp(-Bs^2)$. Noting that function 2 represents I_h/I_{hobs} in some average sense, a plot of the logarithm of function 2 versus s^2 at some average value of s^2 within each group would lead to a straight line graph whose slope would be B and whose intercept would be $-\log k$. Alternatively, it is not necessary to assume a form for the disorder-factor. Function 2 could be plotted directly as a function of s, at some average value of s within each group, and the resulting curve would have $1/k$ as intercept when $s=0$. The shape of the curve gives the statistical positional disorder correction which need not be gaussian. My laboratory has preferred this alternative procedure since we have often found that the temperature effect differs noticeably from a gaussian function.

Once the intensities have been corrected for positional disorder and placed on an absolute scale, it is then a simple calculation to obtain the normalized structure-factor magnitudes, $|E_h|$, that play a key role in direct methods procedures (5) from

$$|E_h|^2 = I_h/\epsilon_h s_2 \qquad (3)$$

where the ϵ_h are numbers that vary with space group and type of reflection. The physical interpretation of the quantities ϵ_h and a method for obtaining them has been given by Wilson (6). It is based upon the effect of individual symmetry elements on the statistics of classes of intensities. The ϵ_h are also determinable from probability methods. They have been defined by Hauptman and Karle (5, *e.g.*) in terms of moment integrals (7)

$$\epsilon_h = [m_2^0(h) + m_0^2(h)]/n \qquad (4)$$

where

$$m_i^k(h) = \int_0^1 \int_0^1 \int_0^1 \xi_j^i n_j^k dxdydz \qquad (5)$$

$$\xi_j^i = \xi^i(x_j, y_j, z_j; h) \qquad (6)$$

$$\eta_j^k = \eta^k(x_j, y_j, z_j; h) \qquad (7)$$

and n is the symmetry number of the space group. The quantities ξ and η are the trigonometric parts of the definition of a structure-factor, F_h, *i.e.*,

$$F_h = \sum_{j=1}^{N/m} f_{jh}(\xi_j + i\eta_j) \qquad (8)$$

and they may be found for the various space groups in International Tables for X-Ray Crystallography (8). Alternative group-theoretical ways of computing the ϵ_h have been described by Stewart and Karle (9) and by Iwasaki and Ito (10). Some clarification concerning the connection between the two calculations has been given by Stewart *et al.* (11).

An early illustration of the procedure for correcting for positional disorder and the placement of experimental intensity data on an absolute scale was presented for the mineral colemanite by Karle, Hauptman and Christ (12). This paper also described the calculation of normalized structure-factors and their application to the determination of phases by use of formulas derived from the initial application of probability theory to the phase problem (5).

Statistics & Direct Methods

In a further study of crystallographic statistics, Wilson (13) found to a good approximation the probability distributions for the intensities in non-centrosymmetric and centrosymmetric crystals. For the former, he obtained

$$_1P(I) = s_2^{-1}\exp(-I/s_2) \qquad (9)$$

and for the latter he obtained

$$_{\bar{1}}P(I) = (2\pi s_2 I)^{-\frac{1}{2}} \exp(-I/2s_2). \qquad (10)$$

Wilson also found the probability distribution for a structure-factor, F, in centrosymmetric crystals

$$_{\bar{1}}P(F) = 2(\pi s_2)^{-\frac{1}{2}} \exp(-F^2/2s_2) \qquad (11)$$

and the probability distribution for a structure-factor magnitude in non-centrosymmetric crystals

$$_1P(|F|) = (2/s_2)|F|\exp(-|F|^2/s_2). \qquad (12)$$

Equations 9-12 suggested that it might be possible to develop statistical tests that distinguish between the presence and absence of symmetry centers. Such tests were developed, for example, by Wilson (13), Howells, Phillips and Rogers (14) and Hauptman and Karle (summarized by Karle (7)). Howells *et al.* developed a test based on the fraction of the reflections, $N(z)$, whose intensities are less than or equal to z, where z is a normalized intensity, $z = I/\langle I \rangle$. The two appropriate functions are

$$_1N(z) = 1 - \exp(-z), \qquad (13)$$

$$_{\bar{1}}N(z) = \mathrm{erf}(z/2)^{\frac{1}{2}}. \qquad (14)$$

These tests and ones based on a variety of statistical properties of normalized structure-factor magnitudes, as shown in Table I, have found much useful appliction.

Table I
Selected statistical properties of centrosymmetric ($\bar{1}$) and non-centrosymmetric (1) reflections

Type	$\langle |E_h| \rangle_h$	$\langle |E_h|^2 \rangle_h$	$\langle ||E_h|^2 - 1| \rangle_h$	Fraction of $|E_h|$ greater than		
				1	2	3
$\bar{1}$	0.798	1.000	0.968	0.32	0.05	0.003
1	0.886	1.000	0.736	0.37	0.018	0.0001

Wilson (13) noted that the tests may fail when certain conditions are not fulfilled. Examples of circumstances in which the distributions of intensities would deviate extensively from those given in Eqs. 9-12 are heavy-atom dominance of the intensities, occupation of special positions in the unit cell and pseudosymmetry. Lipson and Woolfson (15) found that a centrosymmetric structure composed of centrosymmetric molecules gave a more extreme distribution of intensities than normally expected, meaning that those characteristics that distinguish centrosymmetric and non-centrosymmetric distributions were exaggerated in the centrosymmetric crystal. They proposed the name hypercentric for such structures and presented a theory to describe the hypercentric distribution. Hargreaves (16) investigated the effect of heavy atoms on intensity distributions and found that tests for the presence of a center of symmetry could be misleading. The statistics of several hypersymmetric distributions were investigated by Rogers and Wilson (17). Repetition arrangements were considered that involve non-crystallographic symmetry. Statistical criteria for the recognition of the hypersymmetric distributions considered were evaluated. A further study by Wilson (18) considered the effect of having crystallographic symmetry, such as m or mm, present in addition to a crystallographic center of symmetry and parallel repetition of a motif through non-crystallographic symmetry. It was found that the variance appropriate to the hypercentric distribution was reduced to a value closer to that of a normal centric distribution.

Probability distributions were extended to higher order terms for structure-factors for centrosymmetric crystals by Karle and Hauptman (19) and for structure-factor magnitudes for non-centrosymmetric crystals by Hauptman and Karle (20). This could potentially afford a way not only to distinguish between the presence and absence of a center of symmetry but also to detect the presence of certain space groups. Limitations of accuracy, however, and the variations from randomness, $e.g.$, as just described, that characterize crystals have prevented the general application of these higher order terms. Most crystals are not random structures.

One approach to the improvement of phase determining formulas when lack of randomness occurs is to develop corrections to the rational dependence of atoms. A set of numbers $r_j, j=1,...,\nu$ is said to be rationally dependent modulo 1 if there exist ν integers m_j, not all zero, such that

$$\sum_{j=1}^{\nu} m_j r_j = u,$$

where u is an integer and r_j is identified with the atomic coordinates x_j or y_j or z_j. The effect of rationally dependent atoms on phase-determining formulas for centrosymmetric crystals was investigated by Hauptman and Karle (21). The study led to a procedure for reinterpreting and modifying some formulas which is based on the examination of subsets of the experimental data and a subsequent renormalization of the structure factors.

It is important to be aware of the lack of randomness in crystal structures because this characteristic can have a serious effect on the applicability of many phase-determining formulas.

Statistics & Direct Methods

Existence of Solution to Structure Problem

The problem of locating atoms in the unit cell of a crystal can be one of great complexity not only because the composition may be quite intricate but also because the relationship between atomic positions and diffracted intensities presents great difficulties for attempts at direct solution. As is very often the case for complicated problems, the attack on the crystal structure problem has been characterized by the discovery of special properties and the mathematical relationships that they generate which could be developed into practical procedures. The identification of particularly useful relationships and their subtle features took place over a number of years. In addition, the discovery of the valuable relationships was followed by a lengthy period of accommodation between theory and practice during which the practical procedures were developed. The philosophical and practical aspects of structure determination in terms of the accommodation between theory and practice have been discussed by Karle (22,23).

As noted previously, the phases of the x-ray amplitudes scattered from a crystal are lost in the recording of the intensities and, therefore, at first glance it would appear that values for the phases could not be recovered from the intensity data. This was generally thought to be the case, at least, until the end of the 1940's, although there had already been developments in the 1930's that were indications to the contrary. Some were mentioned in the Introduction. Another was the derivation of a function generally known as the Patterson function (24,25),

$$P(r) = \sum_h |F_h|^2 \exp(-2\pi i \mathbf{h} \cdot \mathbf{r}) \tag{15}$$

The maxima of this function, which requires no phase information for its calculation, represent the interatomic vectors in a structure. For simple structures, it was possible to identify a sufficient number of interatomic vectors to permit the determination of atomic coordinates. Once the coordinates were known, phase values could be computed. This demonstrated the existence of a clear path from intensity measurements to phase values and indicated that, at least for very simple structures, phase information is recoverable. It has so far turned out in practice to be more feasible, with all but the simplest structures, to first determine phases and then the atomic positions rather than to determine atomic positions directly from the intensities.

It is straightforward to show that the crystal-structure problem is highly overdetermined when a major fraction of the $CuK\alpha$ sphere of scattering is collected. The electron density distribution, $\rho(\mathbf{r})$, in a crystal may be represented by a three-dimensional Fourier series,

$$\rho(\mathbf{r}) = V^{-1} \sum_h F_h \exp(-2\pi i \mathbf{h} \cdot \mathbf{r}) \tag{16}$$

where the coefficients

$$F_h = |F_h| \exp(i\phi_h) \tag{17}$$

are the crystal structure factors and V is the volume of the unit cell. The Fourier inversion of Eq. 16 followed by the conversion of the integral to the sum of contributions from the N discrete atoms in the unit cell gives for the Fourier coefficient

$$|F_h|\exp(i\phi_h) = \sum_{j=1}^{N} f_{jh}\exp(2\pi i\mathbf{h}\cdot\mathbf{r}_j) \tag{18}$$

Equations (18) form a system of simultaneous equations since a large number of vectors h is considered. The unknown quantities are the phases ϕ_h and the atomic position vectors $\mathbf{r}_j$. The known quantities are the magnitudes of the structure-factors, $|F_h|$, obtainable from experiment and the atomic scattering factors, f_{jh}, which are tabulated. Each equation in the system 18 involving complex quantities is, in fact, two equations, one for the real part and one for the imaginary part. The phases can be eliminated from the equations leaving as unknown quantities those $\mathbf{r}_j$ that represent the independent atomic coordinates in an asymmetric unit for the unit cell. A copper target for X-radiation can provide the values of the intensities, $|F_h|^2$, for as many as 150 independent reflections for each atom in the asymmetric unit of a centrosymmetric crystal, and 75 independent reflections for each atom in the asymmetric unit of a non-centrosymmetric crystal. Since each atom has three positional coordinates, the simultaneous equations 18 would be overdetermined by a factor of as much as about 50 for centrosymmetric crystals and 25 for non-centrosymmetric ones. Although there is some inaccuracy of a few percentages in the values of the $|F_h|^2$ and the f_{jh}, the great overdeterminacy of equations 18 makes these errors unimportant.

An important feature of the overdeterminacy is that it provides the rationale and the motivation for the search for a solution to the problem of determining structures directly from the intensities of diffraction. It further implies that there may exist simple phase-determining relationships that have a high probability of being correct. Herbert Hauptman and I were aware of the overdeterminacy of the crystal structure problem when we began to consider this problem in 1948 and were strongly motivated by it. In 1950, we described the overdeterminacy (26).

Non-negativity and Determinantal Inequalities

It is appropriate to discuss inequality theory here since it not only provided the main mathematical relationships that are currently employed in practical procedures for direct phase determination but also stimulated the development of the probability measures that form an important part of these procedures. It will be seen that the inequalities themselves have probabilistic implications. The inequality relations and their probabilistic implications form the foundation mathematics for direct methods.

The derivation of inequality relationships among the structure factors arose from the non-negativity property of the electron density distributions in unit cells. The concept of non-negativity in structure analysis was first developed for the analysis of molecular structure by use of electron diffraction from gaseous molecules. There was a problem in finding a suitable background intensity, representing the atomic coherent and incoherent scattering, so that the molecular interference intensity could be accurately separated from the total intensity of scattering. Since the Fourier transform of the molecular intensity represents the probability of finding interatomic distances in a molecule, it is apparent that this transform must be non-negative. The shape of the background intensity function was determined by requiring that this smooth function have the property that the corresponding molecular intensity function have a non-negative Fourier sine transform (27,28). The success of this application of non-negativity property led to additional applications. The crystal structure problem was one such application.

Relationships occur among the structure-factors as a consequence of the non-negativity of the electron density distribution in crystals. They take the form of an infinite set of determinantal inequalities of increasing order whose elements are the structure-factors, as described by Karle and Hauptman (29). Previous to the development of the determinantal inequalities, Harker and Kasper (30) derived a system of inequalities, when symmetry is present, that arose from the application of the Schwarz and Cauchy inequalities. On further investigation, it was apparent that the validity of the Harker-Kasper inequalities also depended upon the non-negativity of the electron density distribution.

Although the inequalities having principal application to present methods of phase determination are not to be found explicity among the Harker-Kasper inequalities, the latter inequalities have provided two important philosophic insights. The first is that simple inequality formulas can give useful phase information, as shown by Kasper, Lucht and Harker (31) in the solution of the structure of decaborane. The second concerns the probabilistic characteristics. Gillis (32), for example, observed that the implication of an inequality is still probably correct in those cases when the magnitudes of the structure-factors involved are too small to permit a conclusion to be drawn with certainty. It could, perhaps, be argued that Gillis did not recognize the probabilistic implications because he speculated that the smallness of the structure-factor magnitudes may have been due to thermal effects and used a corresponding function to increase the values of the structure-factor magnitudes in order to apply the inequalities. However, no evidence was given that this was, in fact, a thermal effect and the probabilistic interpretation, which may well be the correct one in this case, remains as an alternative interpretation. It would imply that when an inequality almost, but not quite determines that the phase of a structure-factor has some particular value, it does so with a high probability that the value is correct. This is a matter of great importance because it would mean that the simple inequality relations have probabilistic implications which could extend considerably the range of their validity.

Inequalities

The necessary and sufficient condition for the electron density distribution in a crystal to be non-negative is that an infinite system of determinants involving the structure factor, F_h, be non-negative. An example of such a determinant is (29).

$$D = \begin{vmatrix} F_{000} & F_{-k_1} & F_{-k_2} & \cdots & F_{-h} \\ F_{k_1} & F_{000} & F_{k_1-k_2} & \cdots & F_{k_1-h} \\ F_{k_2} & F_{k_2-k_1} & F_{000} & \cdots & F_{k_2-h} \\ \vdots & \vdots & \vdots & & \vdots \\ F_h & F_{h-k_1} & F_{h-k_2} & \cdots & F_{000} \end{vmatrix} \geq 0 \tag{19}$$

The indices in the first column start with 0,0,0 and are distinct, but are arbitrary otherwise. The indices in the first row are the same as those in the first column, but with opposite signs, and are in the same order. The subscript of the element in the *i*th row and the *j*th column is the sum of the subscripts of the elements of the *i*th row and the first column and the first row and the *j*th column. In forming (19), it is assumed that anomalous dispersion is negligible and therefore $F_h^- = F_h^*$.

Inequalities of successively increasing complexity may be obtained from Ineq. 19. The first three are

$$F_{000} \geq 0 \tag{20}$$

$$|F_{k_i}| \leq F_{000} \tag{21}$$

where Ineq. 21 follows from a second order determinant, and for third order,

$$\begin{vmatrix} F_{000} & F_{-k} & F_{-h} \\ F_k & F_{000} & F_{-h+k} \\ F_h & F_{h-k} & F_{000} \end{vmatrix} \geq 0 \tag{22}$$

The determinantal Ineq. 22 provides a relationship among the structure-factors which is of great significance to direct crystal-structure analysis. However, the main features of this inequlity cannot be readily seen when written in the form of Ineq. 22. In order to understand the meaning of Ineq. 22, it is important to rewrite it in the alternative form given in the 1950 paper of Karle and Hauptman (29).

$$\left| F_j - \frac{F_k F_{h-k}}{F_{000}} \right| \leq \frac{\begin{vmatrix} F_{000} & F_{-k} \\ F_k & F_{000} \end{vmatrix}^{1/2} \begin{vmatrix} F_{000} & F_{-h+k} \\ F_{h-k} & F_{000} \end{vmatrix}^{1/2}}{F_{000}} \tag{23}$$

It is seen that Ineq. 23 is of the form

$$|F_h - \delta| \leq r \tag{24}$$

Determinants of higher order can also be written in the form of Ineq. 24. The interpretation of Ineq. 24 is that the structure-factor F_h is bounded by a circle in the

Statistics & Direct Methods

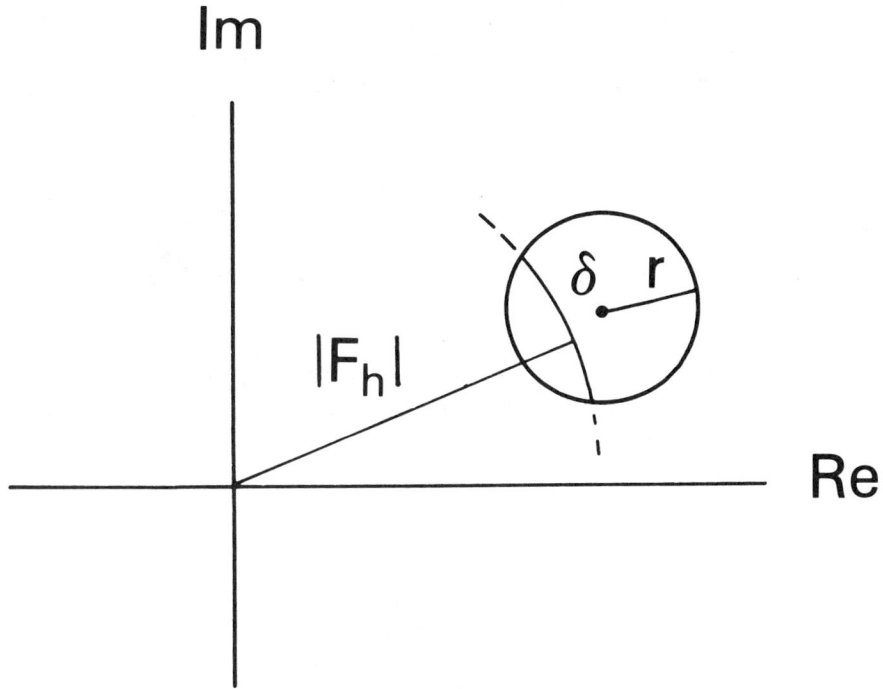

Figure 1. The general form for the determinantal inequalities is $|F_h - \delta| \leq r$. *The diagram shows that this means F_h is bounded by a circle in the complex plane centered at δ and with a radius equal to r. If $|F_h|$ is known, F_h is confined to a line within the circle. If the F's are replaced by U's (unitary structure-factors) and the determinants are of high enough order, r can become vanishingly small or even equal to zero.*

complex plane (Fig. 1) which has its center at δ and a radius r. For Ineq. 23, $\delta = F_k F_{h-k}/F_{000}$ and r is given by the right side of the inequality. As seen from Fig. 1, additional information derives from knowledge of the magnitude $|F_h|$, since the values of F_h are then limited to the segment of the circle of radius $|F_h|$ enclosed within the circle of radius r.

For structure-factors of unusually large magnitude, Ineq. 23 becomes quite restrictive because the right side of Ineq. 23 becomes rather small and it is a simple matter to conclude that ϕ_h, the phase of F_h, is approximately equal to $\phi_k + \phi_{h-k}$, the phase of $F_k F_{h-k}$. By defining "sharpened" structure factors *e.g.* the unitary structure-factors,

$$U_h = F_h / \sum_{j=1}^{N} f_{jh},$$

Ineq. 23 may be replaced by

$$|U_h - U_k U_{h-k}| \leq \begin{vmatrix} 1 & U_{-k} \\ U_k & 1 \end{vmatrix}^{1/2} \begin{vmatrix} 1 & U_{-h+k} \\ U_{h-k} & 1 \end{vmatrix}^{1/2} \quad (25)$$

Such an inequality is strengthened as compared to Ineq. 23 because the right side of Ineq. 25 is relatively smaller in the sense that $|U_k| > |F_k|/F_{000}$ and $|U_{h-k}| > |F_{h-k}|/F_{000}$. It is important at this point to note, however, that the size of the bounding circle is not by itself an adequate measure of the validity of the relation

$$\phi_h \sim \phi_k + \phi_{h-k} \tag{26}$$

Were this the case, Ineqs. 23 and 25 and their implication, the approximate Eq. 26, would appear to be applicable to only the simplest crystals. Probabilistic properties extend the range of applicability of Eq. 26 considerably.

The same reasoning that led to Ineq. 26 from Ineqs. 23 and 25 can lead to other forms of phase-determining formulas, namely, for centrosymmetric crystals,

$$sU_h \sim sU_k U_{h-k} \tag{27}$$

where s means "sign of" and, in general, a tangent formula also follows,

$$\tan\phi_h \simeq \frac{|U_k U_{h-k}|\sin(\phi_k + \phi_{h-k})}{|U_k U_{h-k}|\cos(\phi_k + \phi_{h-k})} \tag{28}$$

In 1952, there were several investigations published that began to reveal the characteristics and value of the sign relationship 27. Sayre (33) derived a formula that is valid when structures contain atoms of essentially equal atomic number (aside from hydrogen atoms),

$$F_h = a_h \sum_k F_k F_{h-k} \tag{29}$$

where a_h is, to a good approximation, a simple function of the scattering angle. He illustrated its use with an application to hydroxyproline. For this simple structure, it was possible to initiate the phase determination by examining Eq. 29 and making some preliminary phase evaluations with little ambiguity. Once the initial evaluations were made, it was possible to proceed with the use of Eq. 29. For more complex substances, however, the initial evaluation of phases must be based on the probabilistic characteristics of formulas such as 27. Influenced by the work of Sayre, Cochran (34) and Zachariasen (5) performed investigations that emphasized the probabilistic features of the sign relationship 27 and its potential for extending the use of that formula.

It is apparent that if **h** is permitted to vary over reciprocal space, Eq. 29 can be considered as a system of simultaneous equations of considerable complexity. Being severely non-linear, there are no general, practical algorithms for obtaining a solution to these equations. Therefore, Eq. 29 cannot be employed in the early stages of a phase determination for any but the simplest of problems. Zachariasen (35) suggested that a Harker-Kasper (30) inequality be used to obtain initial phase values and, once a sufficient set was obtained, that phase extension could proceed with use of

$$s_h = s\langle s_k s_{h-k}\rangle_k \tag{30}$$

where s_h, for example, means the sign of the structure-factor F_h. In this way, he solved the structure of metaboric acid. In doing so, he anticipated several features of direct methods. For example, he recommended that unitary structure-factors of largest magnitude be used in the phase determination. From the previous discussion of the third order inequality and the probability measures that were derived, *e.g.*, with the use of the joint probability distribution (5), it is quite obvious why the largest values of unitary structure-factors or normalized structure-factors give the most reliable indications for phase values.

At present, phase initiation for centrosymmetric crystals proceeds simply by use of relation 27, with the U replaced for convenience by normalized structure-factors. Initial phase values are obtained from specification of the origin in the crystal and some symbolic phases (or signs) associated with normalized structure-factors of largest magnitude. Certain auxiliary formulas can also be used to obtain initial phase values. With the help of appropriate probability measures, the initial set of phases can then be used to effect a structure determination.

My collaborative research program with Herbert Hauptman proceeded on an alternative course that led to the latter procedure. It is apparent, for example, that formula 29 does not provide the means, generally, for supplying an initial set of phases, a main consideration in procedures for phase determination. It is also unlikely that the determination of an initial set of phases by use of inequality theory would be broadly feasible. On the other hand, the third order inequality 23, for example, implied that relations 26, 27, and 28 would follow under favorable circumstances. The problem was to find a way to extend the range of applicability of the inequality relations. Probability theory offered suitable promise in view of the great overdeterminancy of the structure problem. We were quite convinced of the probabilistic character of the determinantal inequality relationshps in general, and had already by 1951 embarked on a program to search out the appropriate formulas. This culminated in the application of the joint probability distribution, at first to centrosymmetric crystals, that gave rise to numerous formulas having probable validity (5) that could be readily identified with formulas in the inequality theory.

Probability measures

After an early attempt to introduce probability methods that made use of the theory of the random walk (36), Herbert Hauptman and I chose the theory of the joint probability distribution for development of phase relationships in centrosymmetric crystals (5). As noted above, our facility with this type of analysis was developed in investigations of intensity statistics for centrosymmetric (19) and non-centrosymmetric crystals (20). Both the joint probability distribution and the central limit theorem have proven to be useful tools in the analysis of statistical properties and the development of phase relationships and probability measures. As will be seen, suitable probability distributions could have been immediately read out of the inequalities written in the form of Ineq. 24 by an application of the central limit theorem, as shown by Karle (37), thus demonstrating another aspect of the inherent probabilistic implications of the inequalities.

The use of the joint probability distribution by Hauptman and Karle (5) produced numerous phase-determining formulas for centrosymemtric crystals having counterparts in the inequality theory, many of which varied with space group, *e.g.*, the sigma-1 and sigma-3 formulas. It also produced probability measures to be applied to the phase-determining formulas, emphasizing the use of normalized structure-factors in phase determination. The theory of origin assignment in crystals was also developed (5) and was expressed in terms of the concept of invariants and seminvariants. Soon after, Woolfson (38) published a probability formula for the sign relation 27 in a convenient hyperbolic tangent form and Cochran (39) published a probability formula for the sum of angles formula 26 for non-centrosymmetric crystals. Both formulas were derived by application of the central limit theorem. In a further application of the joint probability distribution to non-centrosymmetric crystals by Karle and Hauptman (40), several phase-determining formulas for non-centrosymmetric crystals were derived with appropriate probability measures, *e.g.*, the sigma-1 and sigma-2 formulas, further extended to sigma-3 in a subsequent review article (7). In addition, the tangent formula for phase determination and refinement

$$\tan\phi_h \simeq \frac{\sum_h |E_k E_{h-k}|\sin(\phi_k+\phi_{h-k})}{\sum_h |E_k E_{h-k}|\cos(\phi_k+\phi_{h-k})} \qquad (31)$$

was explicitly derived. The tangent formula is also contained in a probability distribution of Cochran (39), as noted by Karle and Karle (41). The counterpart of the tangent formula in determinantal inequality theory is obtained by combining the results of many third order inequalities expressed in the form of relation 28. This work culminated in a practical and efficient procedure for phase determination for centrosymmetric and non-centrosymmetric crystals, the Symbolic Addition Procedure, which is characterized by a stepwise development of phase information by use of simple phase-determining formulas and associated probability measures (41,42).

The probabilistic implications of the inequalities

This section concerns an extension of the previous discussion. Its purpose is to illustrate in some depth the broad inherent probabilistic characteristics of the inequalities and connect them in simple special cases with some of the results of the probability theory discussed above that have found considerable use in the procedures of direct methods.

It is convenient to derive probability formulas from the inequalities in terms of quasi-normalized structure-factors, E,

$$E_h = \left(\sum_{j=1}^{N} f_{jh}^2 \right)^{-1/2} \sum_{j=1}^{N} f_{jh} \exp(2\pi i \mathbf{h} \cdot \mathbf{r}_j) \qquad (32)$$

or

$$E_h \simeq \sigma_2^{-1/2} \sum_{j=1}^{N} Z_j \exp(2\pi i \mathbf{h} \cdot \mathbf{r}_j) \qquad (33)$$

where Z_j is the atomic number of the jth atom and

$$\sigma_n = \sum_{j=1}^{N} Z_j^n \qquad (34)$$

The quasi-normalized structure-factors represent scattering from point atoms and we have for a typical inequality comparable to Ineq. 19 (37),

$$D_{m,p}(h) = \begin{vmatrix} E_{000} & E_{-k_1} & E_{-k_2} & \cdots & E_{-h} \\ E_{k_1} & E_{000} & E_{k_1-k_2} & \cdots & E_{k_1-h} \\ \vdots & \vdots & \vdots & & \vdots \\ E_{k_{m-2}} & E_{k_{m-2}-k_1} & & \cdots & E_{k_{m-2}-h} \\ E_h & E_{h-k_1} & & \cdots & E_{000} \end{vmatrix} \begin{array}{c} \geq 0 \\ (\text{rank } N) \end{array} \qquad (35)$$

The determinant $D_{m,p}(h)$ of order m is formed by composing the first column with E_{000} as the first element followed by $m-2$ arbitrarily chosen structure-factors and then finally E_h. The subscript p labels a particular set of $m-2$ vectors $k_1,...k_{m-2}$ such that $D_{m,p}(h) = D_{m,k_1,...,k_{m-2}}(h)$. In practice, the magnitudes of the structure factors in the first column are generally chosen to be large. Once the first column is specified, the remainder of the determinant is readily constructed in the manner outlined above. It was noted by Goedkoop (43) and Hauptman and Karle (26) that for structure-factors representing point atoms, the determinants of sufficiently high order are equal to zero and therefore become equalities among the structure-factors. The meaning of rank N for Ineq. 35 is that all minors of order $N+1$ or greater are equal to zero, but that at least one minor of order N is different from zero. It was also pointed out by Tsoucaris (44) that the determinants monotonically approach zero as the order of the determinants increases.

The relation comparable to (24) is

$$|E_h - \delta_{m,p}(h)| \leq r_{m,p}(h) \qquad (36)$$

where
$$\delta_{m,p}(h) = \Delta'_{m,p}(h)/\Delta_{m,p} \qquad (37)$$

and
$$r_{m,p}(h) = \Delta^{\frac{1}{2}}_{1,m,p} \Delta^{\frac{1}{2}}_{2,m,p}(h)/\Delta_{m,p} \qquad (38)$$

The determinant Δ' is formed from D by omitting the first row and last column of D, replacing the element E_h by zero and multiplying by $(-1)^{m-1}$,

$$\Delta'_{m,p}(h) = (-1)^{m-1} \begin{vmatrix} E_{k_1} & E_{000} & E_{k_1-k_2} & \cdots & E_{k_1-k_{m-2}} \\ E_{k_2} & E_{k_2-k_1} & E_{000} & \cdots & E_{k_2-k_{m-2}} \\ \vdots & \vdots & \vdots & & \vdots \\ E_{k_{m-2}} & E_{k_{m-2}-k_1} & & \cdots & E_{000} \\ 0 & E_{h-k_1} & & \cdots & E_{h-k_{m-2}} \end{vmatrix} \qquad (39)$$

The determinant Δ is formed from D by omitting the first and last rows and columns of D,

$$\Delta_{m,p} = \begin{vmatrix} E_{k_1-k_1} & E_{k_1-k_2} & \cdots & E_{k_1-k_{m-2}} \\ E_{k_2-k_1} & E_{k_2-k_2} & \cdots & E_{k_2-k_{m-2}} \\ \vdots & \vdots & & \vdots \\ E_{k_{m-2}-k_1} & E_{k_{m-2}-k_2} & \cdots & E_{k_{m-2}-k_{m-2}} \end{vmatrix} \qquad (40)$$

The diagonal elements of Ineq. 40 are evidently E_{000}. The determinants Δ_1 and Δ_2 are formed from D by omitting the last row and column of D and omitting the first row and column of D, respectively,

$$\Delta_{1,m,p}\Delta_{2,m,p}(h) = \begin{vmatrix} E_{000} & E_{-k_1} & \cdots & E_{-k_{m-2}} \\ E_{k_1} & E_{000} & \cdots & E_{k_1-k_{m-2}} \\ E_{k_2} & E_{k_2-k_1} & \cdots & E_{k_2-k_{m-2}} \\ \vdots & \vdots & & \vdots \\ E_{k_{m-2}} & E_{k_{m-2}-k_1} & \cdots & E_{000} \end{vmatrix}$$

$$\times \begin{vmatrix} E_{000} & E_{k_1-k_2} & \cdots & E_{k_1-h} \\ E_{k_2-k_1} & E_{000} & \cdots & E_{k_2-h} \\ E_{k_3-k_1} & E_{k_3-k_2} & \cdots & E_{k_3-h} \\ \vdots & \vdots & & \vdots \\ E_{h-k_1} & E_{h-k_2} & \cdots & E_{000} \end{vmatrix} \qquad (41)$$

Probability measures are directly obtainable from the general inequality 36 that bounds E_h by application of the central limit theorem,

Statistics & Direct Methods

$$p(E) \propto \exp[-(E-\langle E\rangle)^2/2\sigma^2] \tag{42}$$

where $\langle E \rangle$ is the expected value of E and σ^2 is the variance. The quantities $\langle E \rangle$ and σ may be readily associated with quantities in Ineq. 36,

$$\langle E_h \rangle = \delta_{m,p}(h) \tag{43}$$

$$\sigma_{m,p}(h) = \lambda E_{000}^{-1} r_{m,p}(h) \tag{44}$$

where $\lambda = 1$ for centrosymmetric crystals and $\lambda = 2^{-\frac{1}{2}}$ for non-centrosymmetric ones. It is seen that the expected value is associated with the center of the circle that bounds E_h (Fig. 1). The variance is based, quite reasonably, on the radius of the bounding circle. If r is rescaled by multiplying by E_{000}^{-1}, we have the bounding radius when the inequalities are based on unitary structure-factors, i.e. the bounding radius for U_h. The radius is essentially unity when m is small and the magnitudes of the known U are small. When they are large, the radius is significantly smaller than one and it monotonically approaches zero, with the addition of new structure-factor information, as m approaches $N+1$. In other words, certainty is properly associated with determinants that are known to equal zero when all the elements except E_h are known.

If the determinants 35 are composed principally of E's of large magnitude, they approach zero quite rapidly and reach an effective value of zero with increase in order much before the order reaches $N+1$. A general statement can be made to the effect that as the determinants increase in order the inequalities become more restrictive and the rate of increase of restrictiveness with order is greater when the determinants include E's of large magnitude. Current procedures for phase determination are based mainly on third order determinants. It is apparent that a considerable amount of *a priori* phase information is required in order to apply the determinants of high order.

We derive from Ineq. 36 a general form for a phase determining formula which has found widespread application, the tangent formula. We have (37)

$$E_h \propto \langle \delta_{m,p}(h) \rangle_p \tag{45}$$

If we write

$$E_h = |E_h|\exp(i\phi_h) \tag{46}$$

and

$$\delta_{m,p}(h) = |\delta_{m,p}(h)|\exp[i\theta_{m,p}(h)] \tag{47}$$

then the generalized tangent formula follows from expression 45,

$$\tan\phi_h \simeq \frac{\sum_p |\delta_{m,p}(h)|\sin\theta_{m,p}(h)}{\sum_p |\delta_{m,p}(h)|\cos\theta_{m,p}(h)} \qquad (48)$$

It is easily determined from Eq. 37 that the tangent formula presently used corresponds to the case $m=3$. A closely related formula to Eq. 48 for the case $m=3$ is the sum of angles formula Eq. 26 which is also employed to initiate phase determination. For a number of contributions, we have

$$\phi_h \sim \langle \phi_k + \phi_{h-k} \rangle_{k_r} \qquad (49)$$

where k_r means that the average is restricted to ϕ_k and ϕ_{h-k} associated with normalized structure factors of large magnitude. Formula 49 must be used with due regard for the fact that phases are ambiguous modulo 2π (41).

By use of the central limit theorem, expression 42, and the identification of the quantities with those in the general inequality 36 by means of Eqs. 43 and 44, it is possible to associate a probability formula with the general tangent formula 48 and, of course, the special case when $m=3$. The probability formula also applies to Eq. 49. By use of Eqs. 43 and 44, and the definitions 46 and 47, the probability distribution for ϕ_h in terms of other known phases and magnitudes included in various sets of structure-factors, whose vectors are labeled with p, is

$$P(\phi_h) = [2\pi I_0(\alpha)]^{-1}\exp[\alpha\cos(\phi_h - \beta)] \qquad (50)$$

where

$$\alpha = \left\{\left[\sum_p \kappa_{m,p}(h)\cos\theta_{m,p}(h)\right]^2 + \left[\sum_p \kappa_{m,p}(h)\sin\theta_{m,p}(h)\right]^2\right\}^{1/2}, \qquad (51)$$

$$\kappa_{m,p}(h) = 2|E_h \delta_{m,p}(h)|/\sigma^2_{m,p}(h) \qquad (52)$$

and $\tan\beta$ is given by the right side of Eq. 48. Note that this result is based on the assumption that the sets of vectors are sufficiently independent to give essentially independent indications of the value of ϕ_h.

Equation 50 is identical in form with Eq. 3.25 of Karle and Karle (41), and for the case that $m=3$, the equations are the same if we set $\sigma^2_{3,p}(h) = 1$ and replace $E_{000} = \sigma_2^{1/2}/\sigma_1$ with $\sigma_3/\sigma_2^{3/2}$. For equal atoms, both functions become $N^{-1/2}$. Also for the special case of $m=3$ and $\sigma^2_{3,p}(h)=1$, Eq. 50 is equivalent to the probability formulas 6, 7 and 10 of Cochran (39). The variance formula 3.33 and Fig. 2 of Karle and Karle (41) are applicable with the new definition of α, Eq. 51. (A typographical error occurs in Eq. 3.33. The sign before the last term should be minus.)

Statistics & Direct Methods

The evaluation of the variance, $\sigma_{m,p}(h)$, by means of the bounding radius, Eq. 44, in the general inequality 36 represents an advance in the accuracy of the probability formulas. This variance ordinarily occurs with a value of unity in the probability formulas used in procedures for phase determination. When the known structure-factor magnitudes are large and the order of the determinants increases, the variance becomes significantly less than unity, implying that setting it equal to one gives too low a value for the probability. For lower order systems, *e.g.* $m=3$, it may be just as well to maintain the conservative estimate for the probability. This may compensate somewhat for the fact that the data are not perfectly accurate and the "known" phases are not precisely known in a practical application.

For the case of centrosymmetric crystals, the general inequality 36 implies a bound on the real axis with

$$sE_h \simeq s \sum_p \delta_{m,p}(h) \tag{53}$$

where s means 'sign of'. The probability that the sign of E_h is positive can be obtained from application of expression 42 and is found to be

$$P+(h) \simeq \tfrac{1}{2} + \tfrac{1}{2}\tanh |E_h| \sum_p \left[\delta_{m,p}(h) / \sigma^2_{m,p}(h) \right] \tag{54}$$

For $m=3$, and replacement of the quasi-normalized structure-factors by the normalized structure factors E, Eq. 53 becomes the Σ_2 relation (5). Also for $m=3$ and the setting of $\sigma^2_{m,p}(h)=1$, the probability formula 54 corresponds to the one of Cochran and Woolfson (45). It also corresponds to the probability formula 3.30 of Hauptman and Karle (5), as can be more readily seen when formula 3.30 is written in the hyperbolic tangent form (46, p.73).

The determinants have continued to be a fruitful source for probabilistic studies. Conditional determinantal probability distributions have been described by Tsoucaris (44) in which all the elements in the determinants are assumed to be known except those in the last row and column. General determinantal joint probability distributions of essentially the same form as the conditional ones, in which all the elements of the determinants are variates, have been described by Karle (47). Studies concerning the validity of the general determinantal probability distributions have been carried out by Heinerman *et al.* (48,49). The general joint determinantal probability distributions have been used to obtain expected values formulas for evaluating triplet and higher order phase invariants and embedded seminvariants for all the space groups by Karle (50,51,52).

Summary

The study of intensity statistics established procedures for correcting the experimental intensities for positional disorder and scale, led to the concept of the normalized structure-factor and stimulated the first investigations in the joint probability distributions. In a parallel development, formulas that define phase values in terms of measured intensities were derived in the form of inequalities that owe their existence to the non-negativity of the electron density distribution in unit cells. An additional important feature of the structure-factor equations is that the electron density distribution surrounding individual atoms is known to good approximation and therefore the values for atomic scattering factors are known. This knowledge can be employed to show that the number of data normally available from the Cu$K\alpha$ sphere of X-ray scattering makes the crystal-structure problem greatly overdetermined.

Because the crystal-structure problem is greatly overdetermined for small and moderate size structures (not usually for macromolecules), it is reasonable to expect that there exist simple mathematical formulas for defining phases that have a high probability of being correct. The inequality theory presents such simple formulas and, in some instances, they are precisely correct when the inequalities are satisfied. Observations of the behavior of the inequalities when the intensities are not quite large enough to satisfy them and insights derived from the geometric analysis of the determinantal inequalities strongly implied that the usefulness of the inequality relations could be considerably extended by use of probability theory. Thus, overdeterminacy and non-negativity combined to form a path to the discovery of phase-determining formulas having, under suitable circumstances, a high probability of being correct.

Probability measures associated with the inequality theory were derived from applications of the joint probability distribution and also the central limit theorem to the structure-factors. The inherent probabilistic implications of the inequalities were also illustrated somewhat later by the demonstration that probability measures could, in fact, be directly read out of detrimental inequalities, essentially by inspection, with use of the central limit theorem.

The development of theories for intensity statistics, the derivation of the inequality relations and the applications of the joint probability distribution and central limit theorem provided the foundation mathematics for the development of direct methods for phase determination. It could be argued that the foundation mathematics for direct methods could be based solely on applications of the joint probability distribution to the phases and magnitudes of the structure-factors. The sole use of probability theory, however, would eliminate an important source of insight and understanding that is afforded by the inequality theory. The inequalities, based on the non-negativity of the electron density distribution, not only provided considerable motivation for developing the probability functions, but also characterized the types of relations to pursue, afforded deep insights into the reasons why the phase relationships are useful, and, finally, provided their determinantal form to act as the main components of a useful formulation of probability theory. For these reasons, I find it preferable to describe the foundation mathematics of direct methods as arising from the inequality theory and its probabilistic implications which have been developed through the use of appropriate probability methods.

Additional reading relevant to topics in this article may be found in two references concerning crystallographic computing (53,54).

Statistics & Direct Methods

References and Footnotes

1. H. Ott, *Zeitschrift für Kristallographie 66,* 136 (1928).
2. K. Banerjee, *Proceedings of the Royal Society A141,* 188 (1933).
3. M. Avrami, *Physical Review 54,* 300 (1938).
4. A. J. C. Wilson, *Nature 150,* 152 (1942).
5. H. Hauptman and J. Karle, *American Crystallographic Association Monograph No. 3,* Polycrystal Book Service, Western Springs, Illinois (1953).
6. A. J. C. Wilson, *Acta Crystallographica 3,* 258 (1950).
7. J. Karle in *Advances in Chemical Physics 16,* Eds., I. Prigogine and S. A. Rice, Interscience, New York, p. 131 (1969).
8. *International Tables for X-ray Crystallography I,* The Kynoch Press, Birmingham (1965).
9. J. M. Stewart and J. Karle, *Acta Crystallographica A32,* 1005 (1976).
10. H. Iwasaki and T. Ito, *Acta Crystallographica A33,* 227 (1977).
11. J. M. Stewart, J. Karle, H. Iwasaki and T. Ito, *Acta Crystallographica A33,* 519 (1977).
12. J. Karle, H. Hauptman and C. L. Christ, *Acta Crystallographica 11,* 757 (1958).
13. A. J. C. Wilson, *Acta Crystallographica 2,* 318 (1949).
14. E. R. Howells, D. C. Phillips and D. Rogers, *Acta Crystallographica 3,* 210 (1950).
15. H. Lipson and M. M. Woolfson, *Acta Crystallographica 5,* 680 (1952).
16. A. Hargreaves, *Acta Crystallographica 8,* 12 (1955).
17. D. Rogers and A. J. C. Wilson, *Acta Crystallographica 6,* 439 (1953).
18. A. J. C. Wilson, *Acta Crystallographica 9,* 143 (1956).
19. J. Karle and H. Hauptman, *Acta Crystallographica 6,* 131 (1953).
20. H. Hauptman and J. Karle, *Acta Crystallographica 6,* 136 (1953).
21. H. Hauptman and J. Karle, *Acta Crystallographica 12,* 846 (1959).
22. J. Karle, *Proceedings of the National Academy of Sciences, U.S.A. 74,* 4707 (1977).
23. J. Karle, *Proceedings of the National Academy of Sciences U.S.A. 75,* 3540 (1978).
24. A. L. Patterson, *Physical Review 46,* 372 (1935).
25. A. L. Patterson, *Zeitschrift für Kristallographie 90,* 517 (1935).
26. H. Hauptman and J. Karle, *Physical Review 80,* 244 (1950).
27. I. L. Karle and J. Karle, *Journal of Chemical Physics 17,* 1052 (1949).
28. J. Karle and I. L. Karle, *Journal of Chemical Physics 18,* 957 (1950).
29. J. Karle and H. Hauptman, *Acta Crystallographica 3,* 181 (1950).
30. D. Harker and J. S. Kasper, *Acta Crystallographica 1,* 70 (1948).
31. J. S. Kasper, C. M. Lucht and D. Harker, *Acta Crystallographica 3,* 436 (1950).
32. J. Gillis, *Acta Crystallographica 1,* 174 (1948).
33. D. Sayre, *Acta Crystallographica 5,* 60 (1952).
34. W. Cochran, *Acta Crystallographica 5,* 65 (1952).
35. W. H. Zachariasen, *Acta Crystallographica 5,* 68 (1952).
36. H. Hauptman and J. Karle, *Acta Crystallographica 5,* 48 (1952).
37. J. Karle, *Acta Crystallographica B27,* 2063 (1971).
38. M. M. Woolfson, *Acta Crystallographica 7,* 61 (1954).
39. W. Cochran, *Acta Crystallographica 8,* 473 (1955).
40. J. Karle and H. Hauptman, *Acta Crystallographica 9,* 635 (1956).
41. J. Karle and I. L. Karle, *Acta Crystallographica 21,* 849 (1966).
42. I. L. Karle and J. Karle, *Acta Crystallographica 16,* 969 (1963).
43. J. A. Goedkoop, *Acta Crystallographica 3,* 374 (1950).
44. G. Tsoucaris, *Acta Crystallographica A26,* 492 (1970).
45. W. Cochran and M. M. Woolfson, *Acta Crsytallographica 8,* 1 (1955).
46. J. Karle in *Advances in Structure Research by Diffraction Methods 1,* Ed. R. Brill, Interscience, New York, p. 55 (1964).
47. J. Karle, *Proceedings of the National Academy of Sciences U.S.A. 75,* 2545 (1978).
48. J. J. L. Heinerman, H. Krabbendam and J. Kroon, *Acta Crystallographica A35,* 101 (1979).
49. J. J. L. Heinerman, J. Kroon and H. Krabbendam, *Acta Crystallographica A35,* 105 (1979).
50. J. Karle, *Proceedings of the National Academy of Sciences U.S.A. 79,* 1337 (1982).
51. J. Karle, *Proceedings of the National Academy of Sciences U.S.A. 79,* 2125 (1982).

52. J. Karle, *Acta Crystallographica A38,* 327 (1982).
53. *Crystallographic Computing,* Eds., F. R. Ahmed, S. R. Hall and C. P. Huber, Munksgaard, Copenhagen, (1969).
54. *Crystallographic Computing Techniques,* Eds., F. R. Ahmed, K. Huml and B. Sedlacek, Munksgaard, Copenhagen, (1976).

Structure & Statistics in Crystallography, ISBN 0-940030-10-1,
Ed., A. J. C. Wilson F. R. S., Adenine Press, ©Adenine Press 1985.

Fourier Series and Other Representations of Crystallographic pdf's

George H. Weiss[1], Uri Shmueli[2], James E. Kiefer[1], and A. J. C. Wilson[3]

[1]Physical Sciences Laboratory, Division of Computer Research & Technology,
National Institutes of Health, Bethesda, MD 20205 USA
[2]Department of Chemistry, Tel-Aviv University, 69978 Tel-Aviv, Israel
[3]Crystallographic Data Centre, University Chemical Laboratory,
Cambridge CB2 1EW, England

Abstract

The determination of crystal symmetry and the solution of the phase problem often require calculation of univariate or multivariate densities of structure factors. These calculations are almost universally done by using approximations based on the central limit theorem. While such techniques are useful whenever scattering factors are roughly equal, the presence of extreme heterogeneity can lead to a considerable degradation in accuracy. We summarize recent work on exact representations of pdf's in terms of Fourier series. It is shown that the coefficients of these series can be written in terms of the characteristic function evaluated at specified arguments. The resulting univariate series are generally quickly convergent. A table of characteristic functions for many space groups is given.

In addition to these exact results we summarize a technique for calculating the pdf of the structure factor near the maximum value it can take.

1. Introduction

It has long been known that the solution to many problems in the interpretation of crystallographic data can be expressed in terms of statistical properties of two-dimensional random walks first mentioned in a note by Karl Pearson [Pearson (1); Hauptman and Karle (2); Karle and Hauptman (3); Hauptman and Karle (4,5)]. Many techniques for the identification of crystalline structure require a knowledge of the probability density functions (pdf's) of one or more structure factors. Since the 1940's [Wilson (6,7)] these calculations have leaned heavily on the central limit theorem to produce, as a first approximation, a univariate or multivariate Gaussian density. Since the central limit theorem does not produce a satisfactory approximation when extreme heterogeneity exists, especially for space groups of low symmetry, corrections are introduced in terms of moments of the structure factor by means of an expansion in a series of orthogonal polynomials [Klug (8); Karle and Hauptman (9)]. Many variants of this theme exist in the crystallographic literature in which the lowest-order approximation is not a Gaussian but some other density that takes into account relevant crystallographic and non-crystallographic symmetries [Lipson and Woolfson (10); Wilson (11); Rogers and Wilson (12); Wilson (13); Shmueli and Wilson (14,15)]. The effectiveness of these methods is especially enhanced by the availability of symbolic manipulation programs to calculate moments of the structure factor [Shmueli and Wilson (15); Shmueli and Kaldor (16)].

In spite of the heavy dependence of crystallographic analysis on these mathematical approximations there seems to be no systematic examination of their accuracy in the literature. Further, the addition of higher-order terms in orthogonal-polynomial expansions involves algebraic manipulations of increasing complexity. While it is always possible to use simulated data to test accuracy, it is clearly desirable to have mathematically exact representations of the pdf's whenever these can be calculated. For space groups $P1$ and $P\bar{1}$ such representations have been available since the work of Kluyver (17), albeit in a nearly unusable form. Let the structure factor for the space group $P1$ be denoted by $F = A + iB$, where the real and imaginary parts of F are

$$A = \sum_{j=1}^{n} f_j \cos(2\pi \mathbf{h} \cdot \mathbf{r}_j), \quad B = \sum_{j=1}^{n} f_j \sin(2\pi \mathbf{h} \cdot \mathbf{r}_j) \tag{1}$$

where n is the number of atoms in the unit cell. On the assumption that the atoms occupy general positions in the unit cell Kluyver's analysis indicated that the pdf of $|F| = (A^2 + B^2)^{1/2}$ can be written

$$p(|F|) = |F| \int_0^\infty \omega J_0(\omega F) C(\omega) d\omega \tag{2}$$

where $C(\omega)$ is the product

$$C(\omega) = \prod_{j=1}^{n} J_0(f_j \omega). \tag{3}$$

Similarly one can show that in $P\bar{1}$, where $B=0$ by symmetry, the pdf of A is

$$g(A) = \frac{1}{2\pi} \int_{-\infty}^{\infty} e^{-i\omega A} \prod_{j=1}^{n/2} J_0(2f_j \omega) d\omega. \tag{4}$$

Although exact, these results are not too useful for detailed calculations because they require numerical integration of oscillatory integrands, which is generally awkward. Furthermore, Eqs. (2) and (4) give no indication that both $p(|F|)$ and $g(A)$ vanish whenever

$$F^2, A^2 > S_1^2 \tag{5}$$

where

$$S_1 = \sum_{j=1}^{n} f_j \tag{6}$$

2. Fourier Representations of the Centrosymmetric Pdf

In this paper we describe a method for generating exact representations of pdf's of crystallographic random walks in terms of Fourier series. The basic ideas on which this work builds are due to Barakat (18,19) who applied them to the pdf's of models of polymer configuration and laser speckle, and Barakat and Cole (20) who applied them to the statistics of narrow-band noise. The simplest case that illustrates the technique is that of the representation of $g(A)$, defined earlier as the pdf of A in the space group $P\bar{1}$. The important observation that leads to the representation given below is that $g(A)$ can differ from zero only in the interval $(-S_1, S_1)$. Therefore, we can expand $g(A)$ in terms of a Fourier cosine series on that interval:

$$g(A) = \sum_{m=-\infty}^{\infty} a_m \cos\left(\frac{\pi m A}{S_1}\right), \tag{7}$$

since $g(-A) = g(A)$. The a_m are then given by

$$a_m = \frac{1}{2S_1} \int_{-S_1}^{S_1} g(A) \cos\left(\frac{\pi m A}{S_1}\right) dA$$

$$= \frac{1}{2S_1} \int_{-\infty}^{\infty} g(A) \cos\left(\frac{\pi m A}{S_1}\right) dA \tag{8}$$

$$= \frac{1}{2S_1} \int_{-\infty}^{\infty} g(A) \exp\left(\frac{\pi i m A}{S_1}\right) dA.$$

Let us now recall that the characteristic function corresponding to a pdf, $g(A)$, is defined by

$$C(\omega) = \int_{-\infty}^{\infty} p(A) \exp(i\omega A) dA. \tag{9}$$

This allows us to express a_m as

$$a_m = \frac{1}{2S_1} C\left(\frac{\pi m}{S_1}\right) = \frac{1}{2S_1} \prod_{j=1}^{n/2} C_j\left(\frac{2\pi f_j m}{S_1}\right), \tag{10}$$

where $C_j(\omega)$ is the characteristic function corresponding to the jth independent contributing set of scatterers. In general, any set of atoms that are linked by the trigonometric structure factor, possibly modified by non-crystallographic symmetry, contributes a single characteristic function to Eq. (10) [cf. for example, Shmueli and Weiss (21)]. For example, the characteristic function corresponding to $P1$ is that shown in Eq. (3). Notice that, although Eq. (7) resembles Eq. (4), it is much more convenient in practice because it avoids the problem of numerical integration. We have found that convergence of such series, even in the case of extreme heterogeneity, is not a problem in practice. For a crystal corresponding to a $C_{14}U$ asymmetric unit of $P\bar{1}$, no more than 40 terms were required for the evaluation of $g(A)$, and for a smaller degree of heterogeneity even fewer terms suffice. Even quicker convergence can be achieved through the use of convergence acceleration methods [Kiefer and Weiss (22)]. The functional form of $g(A)$ given by Eq. (7) is the same for all the centrosymmetric space groups. The actual expressions for $g(A)$ are obtained by taking account of symmetry by means of the appropriate trigonometric structure factors in the evaluation of the required characteristic functions.

The traditional method for dealing with atomic heterogeneity is to expand the pdf in an Edgeworth or Gram-Charlier series, which make use of the moments to calculate correction terms. This method is not very accurate for extreme heterogeneities, as is suggested by the curves in Fig. 1, which show the exact pdf and approximations to it for a $P\bar{1}$ structure. The solid curve is the pdf of the projection, A, for a molecule in which the asymmetric unit consists of 14 carbon atoms and a single uranium atom. The solid curve is the theoretical result computed from Eq. (7) in both Figs. 1a and 1b, and Fig. 1a shows the comparable approximation by a Gaussian, and an Edgeworth expansion using two moments. It is evident that with zero or two moments the peak seen in the exact curve is not reproduced at all. Figure 1b shows the results obtained when four and eight moments are used to generate corrections. Although these do reproduce the peak, the maximum is displaced from its accurate position. A better approximation is available but involves considerably more effort to produce. If one takes as the lowest-order approximation the pdf produced by the method of steepest descents [Daniels (23); Weiss and Kiefer (24)] and generates polynomials orthogonal with respect to that function, one obtains the results shown in Fig. 2. Although the uncorrected approximation is not a good one, the corrections afforded by the use of two or four moments lead to good qualitative agreement between the exact and approximate pdf's. This is because the relative error in the steepest descent approximation is $O(n^{-3/2})$ rather than the $O(n^{-1/2})$ that is the order of the error given by the central limit theorem [Daniels (25)]. However, it should be noted that considerable computation is required to generate both the lowest-order approximation and the orthogonal polynomials.

3. Fourier and Other Representations of the Noncentrosymmetric Pdf

In the absence of centrosymmetry one is interested in calculating the pdf of the intensity $|F| = (A^2 + B^2)^{1/2}$. At least three approaches to this problem suggest themselves, and are useful in practice although there are restrictions to be noted.

Fourier Series Pdf's

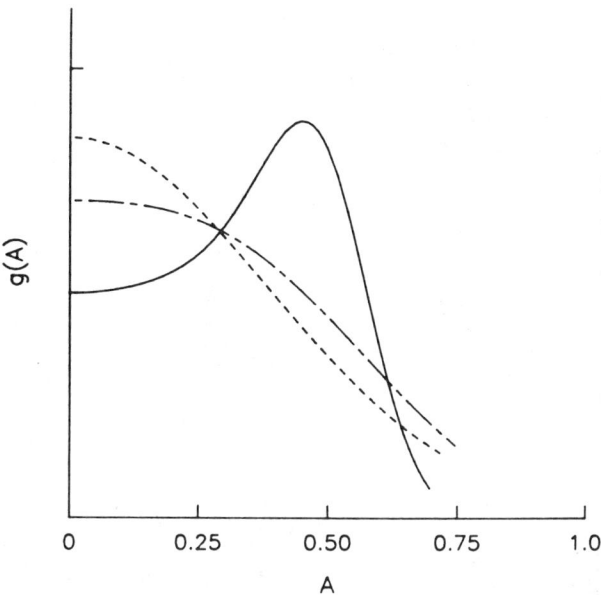

Figure 1a. Curves of the exact and some approximations to the pdf of the projection for a unit cell in $P\bar{1}$, in which the asymmetric unit contains 14 carbon atoms and one uranium atom. Since $g(A) = g(-A)$ only the positive axis is shown. The abscissa is a normalized scale so that $A_{max} = 1$. The exact pdf is shown as a solid line, while the approximate pdf's are an uncorrected Gaussian (-------) and a Gaussian corrected with two moments (—-—-).

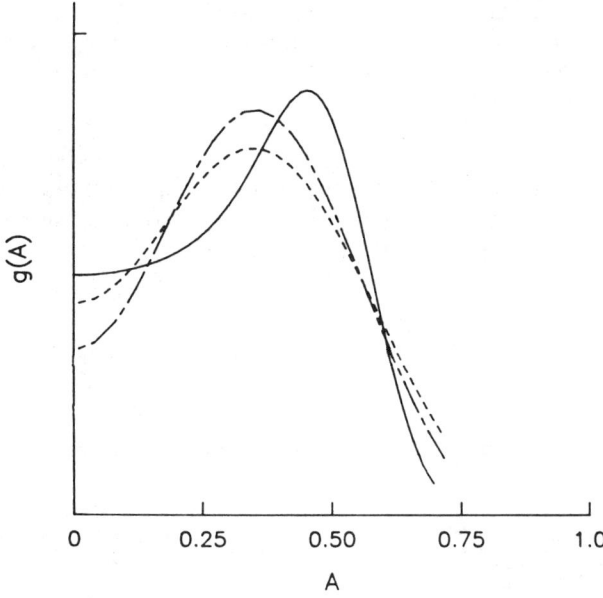

Figure 1b. A comparison of the exact pdf for the same unit cell with the Edgeworth series using 4 (-------) and 8(—-—-) moments.

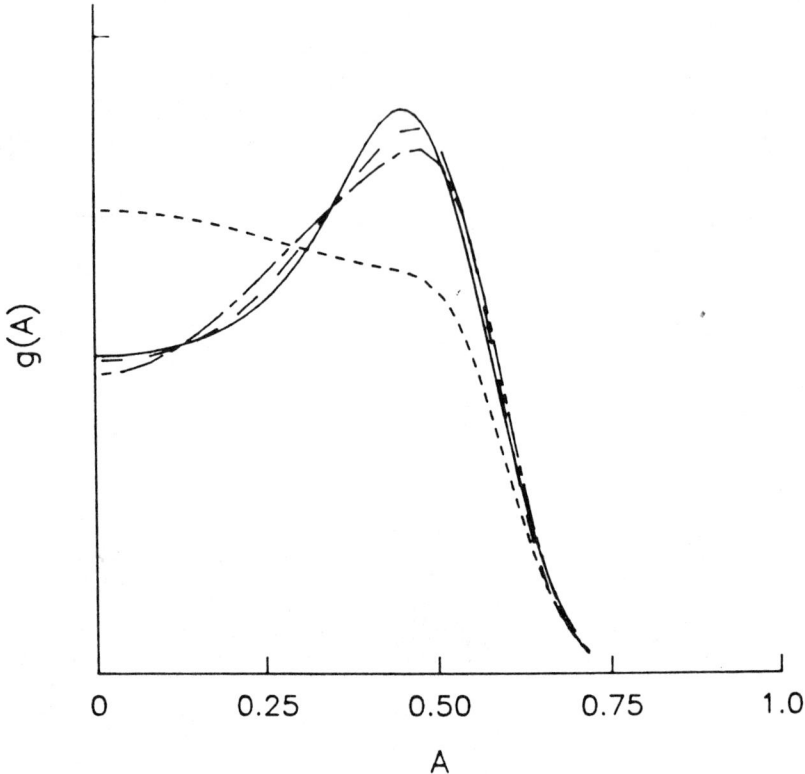

Figure 2. Comparison of the exact pdf with approximations derived from the steepest-descents approximation. The uncorrected steepest-descents approximation is denoted by (········), the two-moment corrected approximation by (—·—·—·), and the four-moment corrected approximation by (— — —). Notice that the peaks appear in their proper location.

The most general approach is directly analogous to Eq. (7) in two dimensions and is valid for any crystal symmetry. If we define the two-dimensional characteristic function of the joint pdf of A and B, $g(A,B)$, to be

$$C(\omega_1,\omega_2) = \int_{-S_1}^{S_1}\!\!\int g(A,B) \exp[i(\omega_1 A + \omega_2 B)]dA dB, \qquad (11)$$

then $g(A,B)$ is expanded in a two-dimensional Fourier series

$$g(A,B) = \frac{1}{4S_1^2} \sum_{r=-\infty}^{\infty} \sum_{s=-\infty}^{\infty} C\left(\frac{\pi r}{S_1}, \frac{\pi s}{S_1}\right) \exp\left[-\frac{\pi i}{S_1}(rA + sB)\right]. \qquad (12)$$

This expression can be converted into a single-series representation for the pdf of $|F|$ by introducing polar coordinates

$$A = |F| \cos \phi, \quad B = |F| \sin \phi \tag{13}$$

into Eq. (12) and integrating over ϕ. The pdf of $|F|$ can then be expressed as $p(|F|)$ where now

$$p(|F|) = |F| \int_0^{2\pi} g(|F| \cos \phi, |F| \sin \phi) d\phi$$

$$= \frac{\pi |F|}{2 S_1^2} \sum_{r=-\infty}^{\infty} \sum_{s=-\infty}^{\infty} C\left(\frac{\pi r}{S_1}, \frac{\pi s}{S_1}\right) J_0\left(\frac{\pi (r^2+s^2)^{1/2} F}{S_1}\right) \tag{14}$$

A second representation useful when $g(A,B)$ has rotational symmetry, so that it is a function of $|F|$ only, is to expand $p(|F|)$ in a Fourier-Bessel function series [Berg and McGregor (26)],

$$p(|F|) = \frac{2|F|}{S_1^2} \sum_{m=1}^{\infty} \frac{1}{J_1^2(\gamma_m)} C\left(\frac{\gamma_m}{S_1}\right) J_0\left(\frac{\gamma_m F}{S_1}\right). \tag{15}$$

In this expansion the γ_m are successive roots of $J_0(\gamma) = 0$, the first five of which are

$$\gamma_1 = 2.4048255577$$
$$\gamma_2 = 5.5200781103$$
$$\gamma_3 = 8.6537279129$$
$$\gamma_4 = 11.7915344390$$
$$\gamma_5 = 14.9309177085.$$

Higher-order zeros can be generated in terms of the variable $\beta_n = (n-\frac{1}{4})\pi$ by [Abramowitz and Stegun (27)]:

$$\gamma_m = \beta_m + \frac{1}{8\beta_m} - \frac{124}{3(8\beta_m)^3} + \frac{120{,}928}{15(8\beta_m)^5} - \frac{401{,}743{,}168}{105(8\beta_m)^7} + \dots \tag{16}$$

The relative error in using this formula is less than 10^{-11} when n exceeds 5. As for the one-dimensional series we have found that the convergence of Eq. (15) presents no difficulties in practice even in the presence of extreme heterogeneity [Shmueli et al. (28)].

The third representation also requires rotational symmetry for its validity [Weiss and Kiefer (29)]. It leads to an expression for $p(|F|)$ in terms of the pdf, $g(A)$, of the projection on an arbitrary axis. One expects such a relation to exist since rotational symmetry implies that the statistical properties of the projection along any axis should be independent of the axis chosen. To derive this relation one substitutes the expression for $C(\omega)$ as a Fourier transform of $g(A)$ into Eq. (2). This leads to the representation

$$p(|F|) = 2|F| \int_0^{S_1} g(A) dA \int_0^\infty \omega J_0(\omega F) \cos \omega A \, d\omega$$

$$= 2|F| \int_0^{S_1} g(A) \{\frac{\partial}{\partial A} \int_0^\infty J_0(\omega F) \sin \omega A \, d\omega\} dA \quad (17)$$

$$= 2|F| \int_0^{S_1} g(A) \frac{\partial}{\partial A} [\frac{H(A-|F|)}{\sqrt{A^2-F^2}}] dA$$

where $H(x)$ is a Heaviside step-function. An integration by parts leads to the final expression

$$p(|F|) = -2|F| \int_{|F|}^\infty \frac{g'(A) dA}{\sqrt{A^2-F^2}}$$

$$= -2|F| \int_0^\infty g'(|F| \cosh u) du. \quad (18)$$

While this relation is not necessarily more convenient than the two earlier ones it does have the virtue of allowing one to generate a useful series for the odd moments of $|F|$. We find

$$\langle |F|^m \rangle = -2 \int_0^\infty |F|^{m+1} d|F| \int_0^\infty g'(|F| \cosh u) du$$

$$= -2 \int_0^\infty \frac{du}{\cosh^{m+2} u} \int_0^\infty v^{m+1} g'(v) dv, \quad (19)$$

where we have interchanged the orders of integration and made the substitution $v = |F| \cosh u$ to get from the first line to the second. If one now integrates the v integral by parts one finds that

$$\langle |F|^m \rangle = 2(m+1) \int_0^\infty \frac{du}{\cosh^{m+2} u} \int_0^\infty v^m g(v) dv$$

$$= 2\sqrt{\pi} \frac{\Gamma(\frac{m+2}{2})}{\Gamma(\frac{m+1}{2})} \int_0^\infty v^m g(v) dv \quad (20)$$

$$= \sqrt{\pi}\, \frac{\Gamma(\frac{m+2}{2})}{\Gamma(\frac{m+1}{2})} \int_0^\infty |v|^m g(v)\,dv \qquad (20)$$

where the absolute value sign has been introduced because $g(A)$ is an even function. Thus we have expressed the moments of $|F|$ in terms of moments of the projection A. When m is an even integer we have the simple relation

$$\langle |F|^{2s} \rangle = 4^s \frac{(s!)^2}{(2s)!} \langle A^{2s} \rangle. \qquad (21)$$

When the order of the moment is odd we can substitute the Fourier series for $g(A)$ into the definition of the absolute moment to find

$$\langle |F|^{2s+1} \rangle = \sqrt{\pi}\, \frac{\Gamma(s+3/2)}{s!} S_1^{2s+1} \Big\{ \frac{1}{2s+1} + \\ + 2 \sum_{j=1}^\infty C\Big(\frac{\pi j}{S_1}\Big) a_{2s+1}(j) \Big\} \qquad (22)$$

where

$$a_{2s+1}(j) = (-1)^{s+1} \frac{(2s+1)!}{(\pi j)^{2s+2}} \Big[1 - (-1)^j \sum_{k=0}^s (-1)^k \frac{(\pi j)^{2k}}{(2k)!} \Big]. \qquad (23)$$

The resulting series in Eq. (22) converges quickly.

It should also be noted that central limit theorem results are implicitly included in both the Fourier and Fourier-Bessel series. Since the Fourier representations are exact it is interesting to examine the assumptions that need to be made to make the transition to the more familiar forms. We will do this for the Fourier series only. In the simplest case of centrosymmetry, in which $g(A)$ can be represented by the series in Eq. (7), Fourier coefficients are expressed as a product of Bessel functions as given in Eq. (3), which may be rewritten as

$$C(\omega) = \exp\Big[\sum_{j=1}^{n/2} \ln J_0(2 f_j \omega) \Big]. \qquad (24)$$

On the assumption that $f_j/S_1 \ll 1$ we can observe that the major contributions to the Fourier series comes from the lowest-order term, which corresponds to small ω in the characteristic function. This suggests the expansion of $C(\omega)$ for small values of ω. Since $J_0(2\omega f_j) \approx 1 - (\omega^2 f_j^2)$, we have

$$C(\omega) \sim \exp\Big[\sum_{j=1}^{n/2} \ln(1-\omega^2 f_j^2) \Big] \\ \sim \exp\Big(-\omega^2 \sum_{j=1}^{n/2} f_j^2 \Big) \qquad (25)$$

in which approximation the expression for $g(A)$ becomes

$$g(A) \sim \frac{1}{2S_1} \sum_{m=-\infty}^{\infty} \exp\left(-\frac{\alpha^2 m^2}{S_1^2}\right) \cos\left(\frac{\pi m A}{S_1}\right) \tag{26}$$

where

$$\alpha = \pi \left(\sum_{j=1}^{n/2} f_j^2\right)^{1/2}. \tag{27}$$

One can now evaluate the sum approximately by replacing it by an integral

$$g(A) \sim \frac{1}{2} \int_{-\infty}^{\infty} \exp(-\alpha^2 x^2) \cos(\pi A x) dx$$

$$= \frac{\sqrt{\pi}}{2\alpha} \exp\left(-\frac{\pi^2 A^2}{4\alpha^2}\right). \tag{28}$$

This is equivalent to the central limit result of Wilson (7). The assumptions that allow us to derive this form of $g(A)$ are that no scattering factor is outstandingly large, and that

$$\frac{\alpha^2}{f^2} = \frac{\pi^2 \sum_{j=1}^{n/2} f_j^2}{\left(\sum_{j=1}^{n/2} f_j\right)^2} \ll 1. \tag{29}$$

Higher-order corrections to Eq. (28) can be derived by properly retaining higher-order terms in the expansion of the Bessel function in Eq. (24). This procedure is equivalent to the more familiar expansion in terms of Hermite polynomials. Although we have given this analysis for $P\bar{1}$ it is readily generalized to other space groups by expanding the characteristic function in terms of moments.

4. Approximation for Large Structure Factors

Wilson (30,31) has raised the question of how to calculate pdf's of large structure factors, *e.g.*, the calculation of $g(A)$ near $A \approx S_1$. An accurate approximation for $g(A)$ in $P\bar{1}$ in this neighborhood was found by Weiss and Kiefer (24), valid for any degree of heterogeneity. In general, the projection A can be written as a sum of scattering factors, which we write

$$A = \sum_j f_j \xi_i(\theta_j). \tag{30}$$

Fourier Series Pdf's

In Eq. (4) we can rotate the axis of integration by setting $\omega = -iv$. This leads to the expression

$$g(A) = \frac{1}{2\pi i} \int_{-i\infty}^{i\infty} e^{-vA} C(iv) dv$$

$$= \frac{1}{2\pi i} \int_{-i\infty}^{i\infty} e^{v(S_1-f)} e^{-vS_1} C(iv) dv \qquad (31)$$

$$= \frac{1}{2\pi i} \int_{-i\infty}^{i\infty} e^{v(S_1-f)} \langle e^{-vf_j(1-\xi_j)} \rangle dv.$$

Thus $g(A)$ can be regarded as an inverse Laplace transform of the function (Widder, 1946)

$$G(v) = \prod_j \langle e^{-vf_j(1-\xi_j(\theta_j))} \rangle. \qquad (32)$$

We may therefore take advantage of the property of the Laplace transform that relates the behavior of $g(A)$ for $A \approx S_1$ to the properties of $G(v)$ for $|v| \to \infty$. This relation is known as a Tauberian theorem in mathematical terminology [Widder (32); Feller (33)]. Specifically, if $G(v)$ behaves like

$$G(v) \sim a v^{-\alpha} \qquad (33)$$

as $|v| \to \infty$ then, subject to some technical conditions that would not be violated for crystallographic problems, it follows that

$$g(A) \sim \frac{a(S_1-A)^{\alpha-1}}{\Gamma(\alpha)} \qquad (34)$$

when A is close to its maximum value. For the case of scattering from a $P\bar{1}$ structure the function $G(v)$ is

$$G(v) = \prod_{j=1}^{n/2} e^{-2vf_j} I_0(2vf_j) \qquad (35)$$

where $I_0(u)$ is a zero'th-order Bessel function of imaginary argument (27). This technique, together with readily computed correction terms, suffices to show that

$$g(A) \sim (4\pi)^{-n/4} \Big(\prod_{j=1}^{n/2} f_j^{-1/2} \Big) \frac{(S_1-A)^{n/4-1}}{\Gamma(n/4)} \Big[1 + \\ + \frac{1}{4(n+4)} \Big(\sum_j f_j^{-1} \Big) (S_1-A) + \dots \Big]. \qquad (36)$$

Detailed calculations with this formula show that the error in the approximation is indistinguishable from zero provided that

$$|S-A| \leq 4 \min_j f_j \qquad (37)$$

[Kiefer and Weiss (22)]. The extension of these results to find the behavior of more general pdf's at large values of the structure factor requires more sophisticated mathematical techniques.

5. Characteristic Functions for Specific Space Groups

We have shown that the pdf's of intensities can be calculated formally in terms of Fourier or Fourier-Bessel series. In Table 1 below we present the components of the coefficients for some of the common space groups. That is to say, if one has a crystal in which the joint characteristic function is

$$C(\omega_1,\omega_2) = \langle e^{i(\omega_1 A + \omega_2 B)} \rangle = \prod_j C_j(\omega_1,\omega_2) \qquad (38)$$

Table I will show the $C_j(\omega_1,\omega_2)$. Three classes of results appear in the table. Those space groups for which $B=0$ so that C_j depends only on ω_1 will be categorized by a I. Those for which ω_1 and ω_2 appear only in the combination $\omega^2 = \omega_1^2 + \omega_2^2$ will be categorized by a II, and those which do not fall into either of the above two categories will be designated as being in class III. Those space groups in category I may be modelled as a one-dimensional random walk whose pdf can be expanded in a Fourier-Bessel series as in Eq. (15), and those in category III require the full two-dimensional expansion as given in Eqs. (12) or (14).

Included in Table I are space groups whose characteristic functions are related to those listed as being the main category. For example, the structure factor in space group $P2$ has the components

$$A = 2 \cos 2\pi(hx + ky)\cos(2\pi lz)$$
$$B = 2 \cos 2\pi(hx + ky)\sin(2\pi lz) \qquad (39)$$

as given in the International Tables [34]. Thus for general reflections

$$C(\omega_1,\omega_2) = \frac{1}{(2\pi)^2} \int\!\!\int_{-\pi}^{\pi} \exp[2if \cos\theta_1(\omega_1 \cos\theta_2 + \omega_2 \sin\theta_2)]d\theta_1 d\theta_2 \qquad (40)$$

$$= \frac{1}{2\pi} \int_{-\pi}^{\pi} J_0(2f\omega \cos\theta)d\theta = J_0^2(\omega f),$$

Fourier Series Pdf's

Table I
Characteristic functions for different space groups. Coefficients of the appropriate Fourier series are expressible as products of these components

Space group	Category	Reflection subset	Characteristic function
$P1$	II	all	$J_0(\omega f)$
$P\bar{1}$	I	all	$J_0(\omega_1 f)$
$P2$	II	hkl	$J_0^2(\omega f)$
	I	$hk0$	$J_0(2\omega_1 f)$
Related Cf's: $P2_1$, $B2$, $C2$, Pb, Pm, Pc, Bm, Cm, Bb, Cc			
$P2/m$	I	hkl	$J_0^2(2\omega_1 f)$
	I	$hk0$	$J_0(4\omega_1 f)$
	I	$00l$	$J_0(4\omega_1 f)$
Related cf's: $P2_1/m$, $B2/m$, $C2/m$, $P2/b$, $P2/c$, $P2_1/b$, $P2_1/c$, $B2/b$, $C2/c$			
$P222$	III	hkl	

$$C(\omega_1,\omega_2) = (2/\pi)^2 \int\!\!\int_0^{\pi/2} J_0[4f(\omega_1^2 \cos^2\theta_1 \cos^2\theta_2 + \omega_2^2 \sin^2\theta_1 \sin^2\theta_2)^{1/2}]d\theta_1 d\theta_2$$

$$= \sum_{s=0}^{\infty}\sum_{t=0}^{\infty}(-1)^{s+t}\frac{(2s)!\,(2t)!}{4^{s+t}[s!t!(s+t)!]^3} f^{2(s+t)}\omega_1^{2s}\omega_2^{2t}$$

Space group	Category	Reflection subset	Characteristic function
	I	$hk0$, $h0l$, $0kl$	$J_0^2(2\omega_1 f)$
	I	$h00$, $0k0$, $00l$	$J_0(4\omega_1 f)$
Related cf's: $P222_1$, $P2_12_12_1$, $C222_1$, $C222$, $F222$, $I222$, $I2_12_12_1$			
$Pmm2$	II	hkl	$\sum_{s=-\infty}^{\infty} J_s^4(2\omega f)$
	I	$hk0$	$J_0^2(2\omega_1 f)$
	III	$h00$	$J_0(4f(\omega_1+\omega_2))$
	II	$0kl$, $h0l$	$J_0^2(\omega f)$
Related cf's: $Pmc2_1$, $Pcc2$, $Pma2$, $Pca2$, $Pnc2$, $Pmn2_1$, $Pba2$, $Pna2$, $Pnn2$, $Cmm2$, $Cmc2_1$, $Ccc2$, $Amm2$, $Abm2$, $Ama2$, $Aba2$, $Fmm2$, $Imm2$, $Iba2$, $Iam2$			
$Pmmm$	I	hkl	$\sum_{s=-\infty}^{\infty} J_s^2(2\omega_1 f)$
	I	$hk0$	$J_0^2(4\omega_1 f)$
Related cf's: $Pnnn$, $Pccm$, $Pban$, $Pmma$, $Pnna$, $Pmna$, $Pcca$, $Pbam$, $Pccn$, $Pbcm$, $Pnnm$, $Pmmn$, $Pbcn$, $Pbca$, $Pnma$, $Cmcm$, $Cmca$, $Cmmm$, $Cccm$, $Cmma$, $Ccca$, $Fmmm$, $Immm$, $Ibam$, $Ibca$, $Imma$			
$P\bar{3}$	I	all	$J_0^3(2\omega_1 f)$
Related cf: $R\bar{3}$			
$P3$	II	all	$J_0^3(\omega f)$
Related cf's: $P3_1$, $P3_2$, $P3$			

while for *hk*0 one finds

$$C(\omega_1) = \frac{1}{2\pi} \int_{-\pi}^{\pi} \exp(2if\omega_1 \cos\theta)d\theta = J_0(2\omega_1 f). \tag{41}$$

A related characteristic function is, for example, *B*2, whose structure factor has the components

$$A = 4\cos^2\frac{\pi(h+l)}{2}\cos 2\pi(hx+ky)\cos(2\pi lz)$$
$$B = 4\cos^2\frac{\pi(h+l)}{2}\cos 2\pi(hx+ky)\sin(2\pi lz) \tag{42}$$

Thus, when $h + l$ is even and $l = 0$ the characteristic function is

$$C(\omega_1,\omega_2) = J_0^2(2\omega f) \tag{43}$$

which is obtainable from Eq. (40).

We point out that Table I is meant to give some representative relations rather than an exhaustive listing. A sample of some of the complications that can arise is furnished by space group *Fdd*2, for which the cases $h + l$ even, $k + l$ even, $h + k + l$ even leads to a characteristic function equivalent to that of *P*222, while $h + k + l$ odd leads to

$$C(\omega) = \frac{4}{\pi^2} \int\!\!\int_0^{\pi/2} J_0(8f\omega\sqrt{\cos^2\theta_1 + \cos^2\theta_2})d\theta_1 d\theta_2$$

$$= \frac{4}{\pi^2} \int\!\!\int_0^{\pi/2} \{J_0(8f\omega\cos\theta_1)J_0(8f\omega\cos\theta_2) +$$

$$+ 2\sum_{l=1}^{\infty}(-1)^l J_{2l}(8f\omega\cos\theta_1)J_{2l}(8f\omega\cos\theta_2)\}d\theta_1 d\theta_2 \tag{44}$$

$$= \sum_{l=-\infty}^{\infty}(-1)^l J_{2l}^4(4f\omega),$$

which resembles, but is not exactly the same as, the characteristic function for *Pmm*2.

Other variants of crystallographic problems can also be fit into the framework of the Fourier-series approach. A simple example of this is the case of the partial disappearance of an inversion center as discussed earlier by Kitaigorodskii (35),

Parthasarathy (36), and Parthasarathy and Parthasarathi (37,38). Of the n atoms in a unit cell let us suppose that m are related by an inversion center and the remaining $n-m$ are uniformly distributed. The structure factor can therefore be expressed as

$$F = 2\sum_{j=1}^{m/2} f_j \cos(2\pi \mathbf{h} \cdot \mathbf{r}_j) + \sum_{j=m+1}^{n} f_j \cos(2\pi \mathbf{h} \cdot \mathbf{r}_j) + \\ + i \sum_{j=m+1}^{n} f_j \sin(2\pi \mathbf{h} \cdot \mathbf{r}_j). \tag{45}$$

Because the structure factor has both a real and an imaginary part a double Fourier series is required as in Eq. (12), and the corresponding joint characteristic function is

$$C(\omega_1,\omega_2) = \prod_{j=1}^{m/2} J_0(2\omega_1 f_j) \prod_{j=m+1}^{n} J_0(\omega f_j) \tag{46}$$

where $\omega^2 = \omega_1^2 + \omega_2^2$.

6. Representation of Joint Pdf's for Direct Methods

Thus far we have developed the formalism appropriate to the study of intensity statistics. An extension of the Fourier method is easily developed for the study of direct methods. These methods are characterized by their use of the joint distribution of several structure factors [Giacovazzo (39)]. Since this subject is so vast we can do no more than sketch some particular analyses, leaving the details with numerical computations to future publications. An almost universal starting point at present for approximating to the joint pdf of several structure factors is to approximate to them by a multivariate Edgeworth or Gram-Charlier series [Klug (8); Hauptman and Karle (40)]. The exact Fourier representations given earlier for the statistical properties of a single intensity are readily generalized to provide representations of joint pdf's. Rather than presenting the most general formalism possible we will consider two particular cases that have appeared in the crystallographic literature.

The first of these involves the joint density of $A(\mathbf{h})$ and $A(2\mathbf{h})$ [Eq. (1)] for space group $P\bar{1}$ [Vand and Pepinsky (41); Cochran and Woolfson (42); Klug (8)]. For simplicity of notation we set $A_1 = A(\mathbf{h})$, $A_2 = A(2\mathbf{h})$, and write $p(A_1,A_2)$ for the sought-after pdf. The formal expression for $p(A_1,A_2)$ as a twofold Fourier series is given in Eq. (12). The individual characteristic functions are

$$C_j(\omega_1,\omega_2) = \langle \exp[2if_j(\omega_1 \cos\theta + \omega_2 \cos 2\theta)] \rangle \\ = \frac{1}{2\pi} \int_{-\pi}^{\pi} \exp[2if_j(\omega_1 \cos\theta + \omega_2 \cos 2\theta)]d\theta. \tag{47}$$

This characteristic function can also be expressed in series form by making use of the expansion

$$\exp(iu \cos \theta) = \sum_{j=-\infty}^{\infty} i^j J_j(u) \exp(ij\theta) \qquad (48)$$

[Gradshteyn and Rzyhik (43)] which can be substituted into the integrand of Eq. (47). When the resulting integrals are evaluated one finds

$$C_j(\omega_1,\omega_2) = \sum_{s=-\infty}^{\infty} i^{3s} J_{2s}(2f_j\omega_1) J_s(2f_j\omega_2) \qquad (49)$$

which has both a real and imaginary part. These are

$$u_j = \operatorname{Re}(C_j) = J_0(2f_j\omega_1)J_0(2f_j\omega_2) + 2 \sum_{m=1}^{\infty} (-1)^m J_{4m}(2f_j\omega_1)J_{2m}(2f_j\omega_2)$$

$$v_j = \operatorname{Im}(C_j) = 2 \sum_{m=0}^{\infty} (-1)^{m+1} J_{4m+2}(2f_j\omega_1)J_{2m+1}(2f_j\omega_2), \qquad (50)$$

which are rapidly convergent. Hence, the characteristic function, $C(\omega_1,\omega_2)$, can be expressed as

$$C(\omega_1,\omega_2) = \prod_j (u_j + iv_j) = U(\omega_1,\omega_2) + iV(\omega_1,\omega_2). \qquad (51)$$

The relevant properties of U and V can be established by examining those of the u_j and v_j. One sees from the exact representations in Eq. (50) that

$$u_j(\pm\omega_1,\pm\omega_2) = u_j(\omega_1,\omega_2)$$

$$v_j(-\omega_1,\omega_2) = v_j(\omega_1,\omega_2) \qquad (52)$$

$$v_j(\omega_1,-\omega_2) = -v_j(\omega_1,\omega_2).$$

Since U is a product of some number of the u_j and an even number of v_j it follows that

$$U(\pm\omega_1,\pm\omega_2) = U(\omega_1,\omega_2). \qquad (53)$$

Further, since V necessarily contains an odd number of v_j, one has

$$V(-\omega_1,\omega_2) = V(\omega_1,\omega_2); \quad V(\omega_1,-\omega_2) = -V(\omega_1,\omega_2). \qquad (54)$$

Since $p(A_1,A_2)$ is necessarily real, the representation in Eq. (12) must reduce to a

Fourier Series Pdf's

combination of sine and cosine series. Using the properties of U and V that we have just established we find that

$$p(A_1, A_2) = \frac{1}{4S_1^2}\left[1 + 2\sum_{s=1}^{\infty}\{U(\tfrac{\pi s}{S_1}, 0)\cos(\tfrac{\pi s A_1}{S_1}) + U(0, \tfrac{\pi s}{S_1})\cos(\tfrac{\pi s A_2}{S_1})\}\right.$$
$$+ 4\sum_{s=1}^{\infty}\sum_{t=1}^{\infty} U(\tfrac{\pi s}{S_1}, \tfrac{\pi t}{S_1})\cos(\tfrac{\pi s A_1}{S_1})\cos(\tfrac{\pi t A_2}{S_1}) + \quad (55)$$
$$\left. + 4\sum_{s=1}^{\infty}\sum_{t=1}^{\infty} V(\tfrac{\pi s}{S_1}, \tfrac{\pi t}{S_1})\cos(\tfrac{\pi s A_1}{S_1})\sin(\tfrac{\pi t A_2}{S_1})\right].$$

The conditional density $p(A_2|A_1)$, analyzed by the authors whose work was cited earlier, is just given by

$$p(A_2|A_1) = p(A_1, A_2)/p(A_1), \quad (56)$$

where $p(A_1)$ is found by integrating $p(A_1, A_2)$ with respect to A_2:

$$p(A_1) = \frac{1}{2S_1}\left[1 + 2\sum_{s=1}^{\infty} U(\tfrac{\pi s}{S_1}, 0)\cos(\tfrac{\pi s A_1}{S_1})\right]$$
$$= \frac{1}{2S_1}\left[1 + 2\sum_{s=1}^{\infty}\{\prod_{j=1}^{n/2} J_0(\tfrac{2\pi f_j s}{S_1})\}\cos(\tfrac{\pi s A_1}{S_1})\right] \quad (57)$$

in agreement with Eq. (7) specialized to the case of $P\bar{1}$. A second result easily obtainable from the expansion in Eq. (55) is $p_+(A_2|A_1)$, the probability that the sign of A_2 is positive, given A_1. This is found from Bertaut's identity (44):

$$p_+ = (1 + p_-/p_+)^{-1} \quad (58)$$

in the form

$$p_+ = \tfrac{1}{2}(1 + \Omega/\Gamma) \quad (59)$$

where

$$\Omega = \frac{1}{S_1^2}\sum_{s=1}^{\infty}\sum_{t=1}^{\infty} V(\tfrac{\pi s}{S_1}, \tfrac{\pi t}{S_1})\cos(\tfrac{\pi s A_1}{S_1})\sin(\tfrac{\pi t |A_2|}{S_1})$$

$$\Gamma = \frac{1}{4S_1^2}\{1 + 2\sum_{s=1}^{\infty}[U(\tfrac{\pi s}{S_1}, 0)\cos(\tfrac{\pi s A_1}{S_1}) + U(0, \tfrac{\pi s}{S_1})\cos(\tfrac{\pi s A_2}{S_1})] + \quad (60)$$
$$+ 4\sum_{s=1}^{\infty}\sum_{t=1}^{\infty} U(\tfrac{\pi s}{S_1}, \tfrac{\pi t}{S_1})\cos(\tfrac{\pi s A_1}{S_1})\cos(\tfrac{\pi t A_2}{S_1})\}.$$

Equation (59) is exact. One can derive the approximation given by Vand and Pepinsky (41) and Cochran and Woolfson (42) by keeping lowest-order terms in ω_1 and ω_2 in Eq. (50). In detail one writes

$$u_j \sim J_0(2f_j\omega_1)J_0(2f_j\omega_2) \sim (1-f_j^2\omega_1^2)(1-f_j^2\omega_2^2)$$
$$\sim 1-f_j^2(\omega_1^2+\omega_2^2), \tag{61}$$
$$v_j \sim -2J_2(2f_j\omega_1)J_1(2f_j\omega_2) \sim -f_j^3\omega_1^2\omega_2.$$

Thus $C_j(\omega_1,\omega_2)$ is, to lowest order,

$$C_j(\omega_1,\omega_2) \sim 1-f_j^2(\omega_1^2+\omega_2^2) - if_j^3\omega_1^2\omega_2, \tag{62}$$

which agrees, again to lowest order, with the exponentiated form

$$C_j(\omega_1,\omega_2) \sim \exp[-f_j^2(\omega_1^2+\omega_2^2) - if_j^3\omega_1^2\omega_2]. \tag{63}$$

In this approximation the characteristic function becomes

$$C(\omega_1,\omega_2) \sim \exp[-\frac{\sigma_2}{2}(\omega_1^2+\omega_2^2) - \frac{\sigma_3}{2}\omega_1^2\omega_2] \tag{64}$$

where

$$\sigma_m = 2\sum_{j=1}^{n/2} f_j^m. \tag{65}$$

The known approximations to $p(A_1,A_2)$ are obtained by further expanding the characteristic function as

$$C(\omega_1,\omega_2) \sim \exp[-\frac{\sigma_2}{2}(\omega_1^2+\omega_2^2)][1-i\frac{\sigma_3}{2}\omega_1^2\omega_2] \tag{66}$$

and pretending that A_1 and A_2 can take on all values between $\pm\infty$. This allows us to use the inversion formula for the Fourier transform

$$p(A_1,A_2) = \frac{1}{4\pi^2}\int\!\!\!\int_{-\infty}^{\infty} C(\omega_1,\omega_2)\exp[-i(\omega_1 A_1+\omega_2 A_2)]d\omega_1 d\omega_2$$
$$\sim \frac{1}{4\pi\sigma_2}\exp[-\frac{1}{4\sigma_2}(A_1^2+A_2^2)]\{1-\frac{\sigma_3}{8\sigma_2^3}A_2(2\sigma_2-A_1^2)\}. \tag{67}$$

Because of all of the approximations inherent in the derivation of this pdf we would expect it to be useful only for a region roughly defined by $A_1^2+A_2^2 < 4\sigma_2$, but to be inaccurate in the tails of the pdf. One can show, from Eq. (67), that

$$p_+ \sim \frac{1}{2}\left(1 + \frac{\sigma_3}{2\sigma_2^{3/2}} |E_2| (E_1^2-1)\right), \qquad (68)$$

as given by Vand and Pepinsky (41) and Cochran and Woolfson (42). While one can derive systematic corrections to this result starting from Eq. (50) the algebra quickly becomes so tedious that there is little point in not using the exact representations of Eq. (55) or (60).

Other examples can readily be solved to furnish exact expressions for the joint pdf. Another case that readily lends itself to Fourier representation is the joint pdf of E_h, E_k, E_{h+k} in $P\bar{1}$ [Klug (8)]. The components of the characteristic function can be expressed either as a double integral or as an infinite series

$$C_j(\omega_1,\omega_2,\omega_3) = u_j + iv_j =$$

$$= \frac{1}{(2\pi)^2} \int\int_{-\pi}^{\pi} \exp[2if_j(\omega_1 \cos\theta + \omega_2 \cos\phi + \omega_3 \cos(\theta + \phi))]d\theta d\phi \qquad (69)$$

in which

$$u_j = J_0(2f_j\omega_1)J_0(2f_j\omega_2)J_0(2f_j\omega_3) +$$

$$+ 2\sum_{s=1}^{\infty}(-1)^s J_{2s}(2f_j\omega_1)J_{2s}(2f_j\omega_2)J_{2s}(2f_j\omega_3), \qquad (70)$$

$$v_j = 2\sum_{s=0}^{\infty}(-1)^{s+1} J_{2s+1}(2f_j\omega_1)J_{2s+1}(2f_j\omega_2)J_{2s+1}(2f_j\omega_3).$$

One can derive an approximation analogous to Eq. (64) by following the same steps as well as the lowest-order approximation for p_+ similar to Eq. (68). A more complete numerical examination of the consequences of these exact representations will be presented in fuller detail elsewhere. However, our preliminary calculations suggest that when extreme atomic heterogeneities occur approximations that are based on the perturbation of an underlying Gaussian pdf can be quite inaccurate. Although the calculation of Fourier coefficients can appear to be a formidable task the series that we have derived in Eqs. (50) and (70) are rapidly convergent and should lead to no difficulties in practice. These matters will be discussed more fully in analyses that are in progress.

References and Footnotes

1. Pearson, K. *Nature 72*, 294 (1905).
2. Hauptman, H. & Karle, J. *Acta Cryst. 5*, 48-59 (1952).

3. Karle, J. & Hauptman, H. *Acta Cryst. 6,* 131-135 (1953).
4. Hauptman, H. & Karle, J. *Acta Cryst. 6,* 136-141 (1953).
5. Hauptman, H. & Karle, J. *Solution of the Phase Problem 1. The Centrosymmetric Crystal,* ACA Monograph No. 3. Pittsburgh: Polycrystal Book Service. (1953b).
6. Wilson, A.J.C. *Nature 150,* 151-152 (1942).
7. Wilson, A.J.C. *Acta Cryst. 2,* 318-321 (1949).
8. Klug, A. *Acta Cryst. 11,* 515-543 (1958).
9. Karle, J. & Hauptman, H. *Acta Cryst. 11,* 264-269 (1958).
10. Lipson, H. & Woolfson, M.M. *Acta Cryst. 5,* 680-682 (1952)
11. Wilson, A.J.C. *Acta Cryst. 5,* 318-324 (1952).
12. Rogers, D. & Wilson, A.J.C. *Acta Cryst. 6,* 439-449 (1953).
13. Wilson, A.J.C. *Acta Cryst. 9,* 143-144 (1956).
14. Shmueli, U. & Wilson, A.J.C. *Acta Cryst. A37,* 342-353 (1981).
15. Shmueli, U. & Wilson, A.J.C. *Acta Cryst. A39,* 225-233 (1983).
16. Shmueli, U. & Kaldor, U. *Acta Cryst. A39,* 619-621 (1983).
17. Kluyver, J.C. *Kon. Akad. van Wet. Amst. 8,* 341-350 (1906).
18. Barakat, R. *J. Phys. A6,* 796-804 (1973).
19. Barakat, R. *Opt. Act. 21,* 903-921 (1974).
20. Barakat, R. & Cole, J.E. *J. Sound & Vib. 62,* 365-377 (1979).
21. **Shmueli, U. & Weiss, G.H.** *this volume,* pp. 53-66.
22. Kiefer, J. E. & Weiss, G.H. *AIP Conference Proceedings No. 109,* 11-32 (1984).
23. Daniels, H.E. *Ann. Math. Stat. 25,* 631-650 (1954).
24. Weiss, G.H. & Kiefer, J.E. *J. Phys. A16,* 489-495 (1983).
25. Daniels, H.E. *Biometrika 43,* 169-185 (1956).
26. Berg, P.W. & McGregor, J.L. *Elementary Partial Differential Equations,* San Francisco: Holden-Day (1966).
27. Abramowitz, M. & Stegun, I.A. *Handbook of Mathematical Functions.* Washington, D.C.: U.S. Government Printing Office (1970).
28. Shmueli, U., Weiss, G.H., Kiefer, J.E., & Wilson, A.J.C. *Acta Cryst. A40,* 551-560 (1984).
29. Weiss, G.H. & Kiefer, J.E. *J. Sound and Vib.* (to appear).
30. Wilson, A.J.C. *Technometrics 22,* 629-630 (1980).
31. Wilson, A.J.C. *Acta Cryst. A39,* 26-28 (1983).
32. Widder, D.V. *The Laplace Transform.* Princeton, N.J.: Princeton University Press (1946).
33. Feller, W. *An Introduction to Probability Theory and its Applications. Vol. 2,* New York: John Wiley and Sons (1970).
34. Henry, N.F.M. & Lonsdale, K. *International Tables for X-Ray Crystallography.* Birmingham: Kynoch Press (1952).
35. Kitaigorodskii, A.I. *The Theory of Crystal Structure Analysis.* New York: Consultants Bureau (1961).
36. Parthasarathy, S. *Z. Kristallogr 123,* 77-80 (1966).
37. Parthasarathy, S., *Acta Cryst. 21,* 642-647 (1966).
38. Parthasarathy, S. & Parthasarathi, V. *Acta Cryst. A32,* 57-59 (1976).
39. Giacovazzo, C. *Direct Methods in Crystallography,* New York: Academic Press (1980).
40. Hauptman, H. & Karle, J. *Acta Cryst. 11,* 149-157 (1958).
41. Vand, V. & Pepinsky, R. *The Statistical Approach to X-Ray Structure Analysis,* State College: Pennsylvania State Press (1953).
42. Cochran, W. & Woolfson, M.M. *Acta Cryst. 7,* 450-455 (1954).
43. Gradshteyn, I.S. & Ryzhik, I.M. *Tables of Integrals, Series, and Products,* New York: Academic Press (1980).
44. Bertaut, E.F. *Acta Cryst. 8,* 823-832 (1955).

Structure & Statistics in Crystallography, ISBN 0-940030-10-1,
Ed., A. J. C. Wilson F. R. S., Adenine Press, ©Adenine Press 1985.

Effect of Heavy Atoms on Intensity Statistics of X-Ray Diffraction by Crystals

D. Pradhan, S. Ghosh and G.D. Nigam
Department of Physics, Indian Institute of Technology
Kharagpur 721 302, India

Abstract

Heavy atoms in fixed special positions are accompanied by excluded volume which is inaccessible to the light atoms in the unit cell. The shape transform of this volume modulates the expected value of intensity in reciprocal space. The effects of excluded volumes around heavy atoms in fixed special positions have been investigated on cumulative distribution functions in space groups $P1$ and $P\bar{1}$. These distribution functions have been compared with corresponding distributions based on three known structures. A reasonably good agreement is found in all cases.

Introduction

It is well known that if the unit cell of a crystal contains one or few heavy atoms, the distributions of X-ray intensities usually deviate considerably from those inferred by the Wilson (1) statistics. In such cases, the resolution of space group ambiguities with the aid of these statistics becomes almost impossible. The effect of heavy atoms on the theoretical $N(z)$ distributions has been first considered by Collin (2) and later by Sim (3) and Srinivasan (4). Sim (3) has considered the case of a single heavy atom at the origin and a large number of light atoms in a triclinic cell. He defined the random variable as

$$\mathcal{T} = F - f_H \qquad (1)$$

which is assumed to be normally distributed with distribution parameters

$$\langle \mathcal{T} \rangle = 0 \qquad (2)$$

and

$$\text{var}(\mathcal{T}) = \Sigma_L \qquad (3)$$

where f_H is the scattering factor of the heavy atom, F the structure factor and Σ_L is the sum of the squares of the scattering factors over the light atoms.

While defining the random variable, the contribution of heavy atom has been subtracted. In doing so, the position of heavy atom has been made inaccessible to the light atoms. Furthermore light atoms cannot come close to the heavy atom by a minimum distance 'a' equal to sum of heavy atom radius and the average radius of the light atoms. This results in a strong correlation between atomic positions, so that the heavy atom is surrounded by an approximately spherical volume, being inaccessible to the light atoms. The shape transform of this volume modulates the average intensity as a function of s. As a result of this, equation (3) is no longer strictly true; a result which has been reached by French and Wilson (5) on empirical grounds and justified theoretically by Wilson (6). The situation is closely analogous to the inaccesible volumes caused by certain symmetry elements (7,8,9). It is also similar to the effect observable in Patterson representation of the actual structure, where a pseudo-atom at the origin is surrounded by an spherical inaccessible volume of one atomic diameter (6).

The object of this paper is to investigate the effect of the excluded volume on the cumulative distribution functions of the normalised intensities when the heavy atom occupies the origin of the unit cell in space groups $P1$ and $P\bar{1}$. A numerical analysis has been made to study their behaviour on the parameters $r (r = f_H/\Sigma_L^{1/2})$ and A and B, occurring in the expressions of $_1N(z, r, A)$ and $_{\bar{1}}N(z, r, B)$. The parameters A and B are structure dependent as they are functions of shape and size of the excluded volume. Values of $_1N(z, r, A)$ $_{\bar{1}}N(z, r, B)$ are listed in Table 1 and Table 2 from the numerical study. The $N(z)$ tests are applied to the data from three known structures and in all cases the correct space group is indicated by the theory.

Derivation of cumulative distribution function

The cumulative distribution functions will now be derived for the space groups $P1$ and $P\bar{1}$, containing one heavy atom and m number of light atoms in terms of parameters r and a, the radius of the spherical excluded volume in the unit cell.

Space group P1

Let the origin be chosen on the heavy atom. The structure factor is

$$F = f_H + \sum_{i=1}^{m} f_i \exp 2\pi i s \cdot r_i \tag{4}$$

$$\text{or } \mathscr{T} = F - f_H = \sum_{i=1}^{m} f_i \exp 2\pi i s \cdot r_i.$$

It is readily seen that

$$\langle \mathscr{T} \rangle = 0, \tag{5}$$

Heavy Atoms & Intensity Statistics

$$\mathcal{T}\mathcal{T}^* = (F-f_H)(F-f_H)^* = \Sigma_L + 2\sum_{i,j}^{m} f_i f_j^* \cos 2\pi \mathbf{s} \cdot (\mathbf{r}_i - \mathbf{r}_j). \tag{6}$$

It now remains to calculate the expected value of $\mathcal{T}\mathcal{T}^*$ in the restricted volume of the unit cell. The light atoms, instead of being distributed over the whole cell of volume V, are excluded from the spherical volume of radius a surrounding the origin. The volume available to them is $V - (4/3)\pi a^3$, as there is only one excluded volume per unit cell. Integration over the restricted volume needed in the evaluation of the expected value of $\mathcal{T}^*$ is equivalent to integrating first over the whole cell and then subtracting the integral over the interior of the excluded volume. Dispersion, implicit in the use of complex conjugate of f_j in (6), will be neglected. The expected value $\langle \mathcal{T}^* \rangle$ is

$$\langle \mathcal{T}^* \rangle = \Sigma_L - (\Phi - \Sigma_L)A \tag{7}$$

where

$$\Phi = \left(\sum_{i=1}^{m} f_i\right)^2 \tag{8}$$

and

$$A = \frac{36}{(3V - 4\pi a^3)^2 s^2} \left[\frac{\sin(2\pi s a) - 2\pi s a \cos(2\pi s a)}{(2\pi s)^2}\right]^2. \tag{9}$$

In the Appendix, we solve the integral which permits the derivation of (9). From (4), (5), (6) and (7), we get

$$\langle F \rangle = f_H, \tag{10}$$

$$\text{var}(F) = \Sigma_L - (\Phi - \Sigma_L)A. \tag{11}$$

The probability that $|F|$ lies between $|F|$ and $|F| + d|F|$ can readily be written as

$$_1P(|F|)d|F| = \frac{2|F|}{\Sigma_L - (\Phi - \Sigma_L)A} \exp\left[-\left(\frac{|F|^2 + f_H^2}{\Sigma_L - (\Phi - \Sigma_L)A}\right)\right] \\ \times I_0\left(\frac{2|F|f_H}{\Sigma_L - (\Phi - \Sigma_L)A}\right) d|F|. \tag{12}$$

where $I_0(x)$ is a modified Bessel function of zero order.

Now

$$z = |F|^2 / \{\Sigma_L + f_H^2 - (\Phi - \Sigma_L)A\}$$

we get

$$_1P(z,r,A)dz = \frac{[1+r^2-(\Phi/\Sigma_L-1)A]}{1-(\Phi/\Sigma_L-1)A} \exp[-z\{(1+r^2)-(\Phi/\Sigma_L-1)A+r^2\}]$$

$$\times I_0\left[\frac{2z^{1/2}r\{(1+r^2)-(\Phi/\Sigma_L-1)A\}^{1/2}}{1-(\Phi/\Sigma_L-1)A}\right]dz. \tag{13}$$

The cumulative distribution function of z is obtained by evaluating the integral $\int_0^z {}_1P(z,r,A)dz$, which leads to

$$_1N(z,r,A) = \frac{[(1+r^2)-(\Phi/\Sigma_L-1)A]}{1-(\Phi/\Sigma_L-1)} \exp(-r^2) \times$$

$$\times \int_0^z \exp[-z\{(1+r^2)-(\Phi/\Sigma_L-1)A\}]I_0\left[\frac{2z^{1/2}r\{(1+r^2)-(\Phi/\Sigma_L-1)A\}^{1/2}}{1-(\Phi/\Sigma_L-1)A}\right]dz. \tag{14}$$

Space group P$\bar{1}$

$$\mathcal{T} = F - f_H = 2\sum_{i=1}^{m/2} f_i \cos(2\pi \mathbf{s}\cdot\mathbf{r}_i) \tag{15}$$

and

$$\mathcal{T}^2 = (F - f_H)^2 = \Sigma_L + 2\sum_{i=1}^{m/2} f_i^2 \cos 4\pi \mathbf{s}\cdot\mathbf{r}_i$$

$$+ 4\sum_{i,j}^{m/2} f_i f_j \cos(2\pi \mathbf{s}\cdot\mathbf{r}_i) \cos(2\pi \mathbf{s}\cdot\mathbf{r}_j). \tag{16}$$

There are eight excluded volumes at (000), (½00), (0½0), (00½), (½½0), (½0½), (0½½) and (½½½). The volume available for random distribution to the light atoms is $V - 32\pi a^3/3$. In our example, the heavy atom occupies the origin. The excluded volume at the origin is larger than the one at any other centre of inversion. They have been assumed to be of the same size for simplicity.

In evaluating the expected value of $\mathcal{T}^2$ from (16), the average value of first term is Σ_L. In case of h,k,l nonzero, the integral of third term is zero in the whole unit cell as well as in the excluded volume. The middle term, however, is finite within the excluded volume. It has the same value, since $\cos(4\pi \mathbf{s}\cdot\mathbf{r})$ is invariant by a translation of ½00 *etc*. The expected value $\langle \mathcal{T}^2 \rangle$ is

Heavy Atoms & Intensity Statistics

$$\langle \mathscr{F}^2 \rangle = \Sigma_L - B\Sigma_L \tag{17}$$

where

$$B = \frac{24}{(3V - 32\pi a^3)s} \left[\frac{\sin(4\pi sa) - 4\pi sa \cos(4\pi sa)}{(4\pi s)^2} \right]. \tag{18}$$

The integral which permits the derivation of (18) is given in the Appendix. From (10), (16) and (17) we get

$$\mathrm{var}(F) = \Sigma_L - B\Sigma_L. \tag{19}$$

The probability that $|F|$ lies between $|F|$ and $|F|+d|F|$ is

$$_{\bar{1}}P(|F|)d|F| = [2\pi(1-B)\Sigma_L]^{-\frac{1}{2}} \left[\exp - \frac{(|F|-f_H)^2}{2(1-B)\Sigma_L} \right.$$
$$\left. + \exp\{-\frac{(|F|+f_H)^2}{2(1-B)\Sigma_L}\} \right] d|F|. \tag{20}$$

Substituting $z = |F|^2/(1-B)\Sigma_L + f_H^2$, and integrating we get

$$_{\bar{1}}N(z,r,B) = \Psi[(z^{\frac{1}{2}}\sqrt{1-B+r^2} - r)/\sqrt{1-B}]$$
$$+ \Psi[(z^{\frac{1}{2}}\sqrt{1-B+r^2} + r)/\sqrt{1-B}] \tag{21}$$

where

$$\Psi(t) = (2\pi)^{-\frac{1}{2}} \int_0^t \exp(-\frac{1}{2}x^2) dx. \tag{22}$$

Application and Discussion

A computer program was written for the evaluation of distribution functions from (14) and (21). Values of the cumulative distribution functions obtained by numerical computation are presented in tabular form. Table I shows the behaviour of $_1N(z,r,A)$ and $_{\bar{1}}N(z,r,B)$ as a function of r for a constant value of A and B. Table II shows $_1N(z,r,A)$ versus A and $_{\bar{1}}N(z,r,B)$ versus B for a constant value of r. We note that the difference between the cumulative distribution functions is quite significant for large A and B values to distinguish between $P1$ and $P\bar{1}$. It may further be observed that Sim's distributions (3) are the special case of (14) and (21) by letting A and B to be zero.

The modified cumulative distribution functions were compared with the experimental distribution $N(z)$ for rubidium hydrogen di-*o*-nitrobenzoate (3), 3-bromo-

Table I
Values of $_1N(z,r,A)$ and $_{\bar{1}}N(z,r,B)$

z/r	$_1N(z,r,0.08)$				$_{\bar{1}}N(z,r,0.45)$			
	0	1	2	3	0	1	2	3
0.0	0.000	0.000	0.000	0.000	0.000	0.000	0.000	0.000
0.1	0.095	0.082	0.017	0.001	0.248	0.149	0.018	0.000
0.2	0.182	0.163	0.049	0.010	0.345	0.224	0.049	0.005
0.3	0.261	0.240	0.096	0.039	0.416	0.288	0.093	0.020
0.4	0.334	0.313	0.159	0.102	0.472	0.346	0.148	0.049
0.5	0.399	0.381	0.236	0.213	0.520	0.400	0.212	0.098
0.6	0.459	0.444	0.324	0.379	0.561	0.449	0.281	0.166
0.7	0.514	0.503	0.421	0.599	0.597	0.495	0.353	0.250
0.8	0.564	0.557	0.522	0.766	0.628	0.538	0.425	0.345
0.9	0.609	0.607	0.624	0.856	0.657	0.578	0.495	0.444
1.0	0.659	0.653	0.725	0.989	0.682	0.615	0.562	0.542

1,8-dimethylnaphthalene (10) and anorthite (11). The parameters r, A and B vary with the Bragg angle θ, and a suitable average must be used. The values of r at $\sin \theta = 0.45$ were used for the above cases. The values of A and B were evaluated in the following manner. The trigonometric part occurring in A was plotted against sa, which showed marked oscillations with decreasing amplitude. It is mainly the first peak which contributed to the value of A. The radius of the spherical excluded volume was computed from the known atomic radii (12). Knowing a, the value of s was found from the graph where the peak height was half its value at sa

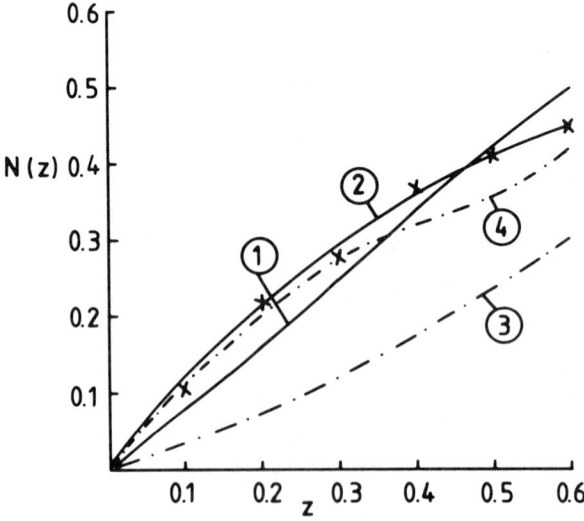

Figure 1. Comparison of experimental $N(z)$ points (marked by crosses) of rubidium hydrogen-o-dinitrobenzoate with theoretical distributions ① and ② as calculated from (14) and (21) for space groups $P1$ and $P\bar{1}$ respectively. The dashed curves ③ and ④ are Sim's plots.

Heavy Atoms & Intensity Statistics

Table II
Values of $_1N(z,r,A)$ and $_{\bar{1}}N(z,r,B)$

z/A	$_1N(z,0.97,A)$ 0.0	0.1	0.2	0.3	z/B	$_{\bar{1}}N(z,0.97,B)$ 0.0	0.1	0.2	0.3
0.0	0.000	0.000	0.000	0.000	0.0	0.000	0.000	0.000	0.000
0.1	0.079	0.086	0.094	0.140	0.1	0.219	0.214	0.207	0.199
0.2	0.156	0.171	0.190	0.215	0.2	0.308	0.302	0.294	0.284
0.3	0.228	0.253	0.284	0.329	0.3	0.376	0.370	0.361	0.351
0.4	0.295	0.330	0.377	0.445	0.4	0.432	0.426	0.417	0.407
0.5	0.355	0.402	0.466	0.561	0.5	0.480	0.474	0.466	0.456
0.6	0.411	0.470	0.552	0.677	0.6	0.523	0.517	0.510	0.501
0.7	0.466	0.533	0.635	0.792	0.7	0.561	0.555	0.549	0.541
0.8	0.517	0.592	0.714	0.905	0.8	0.595	0.590	0.584	0.577
0.9	0.565	0.646	0.789	0.955	0.9	0.626	0.622	0.617	0.610
1.0	0.609	0.696	0.859	0.997	1.0	0.654	0.651	0.646	0.641

$= 0$. The value of A was computed at this particular value of s. Same procedure was followed for the evaluation of B. Figs. 1, 2 and 3 show the results, where the experimental distributions $N(z)$ are compared with the theoretical distributions. The results are in reasonable agreement with the assignment of $P\bar{1}$ as the space group in each of the compounds.

The examples presented here clearly show that the cumulative distribution functions in (14) and (21) are of importance in detecting a centre of inversion when

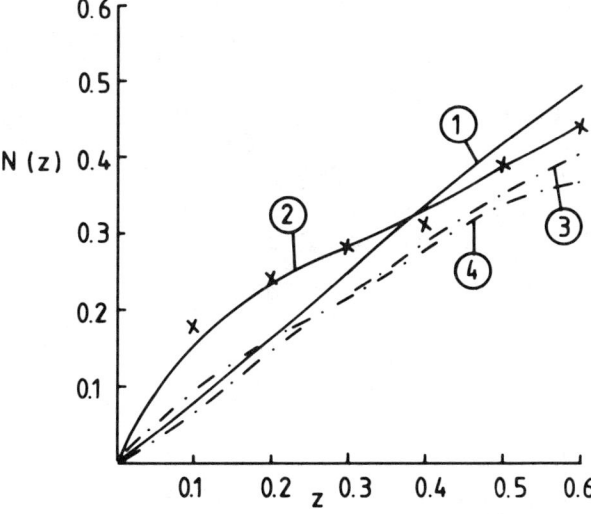

Figure 2. Comparison of experimental $N(z)$ points (marked by crosses) of 3-bromo-1,8 dimethyl-naphthalene with theoretical distributions ① and ② as calculated from (14) and (21) for space groups $P1$ and $P\bar{1}$ respectively. The dashed curves ③ and ④ are Sim's plots.

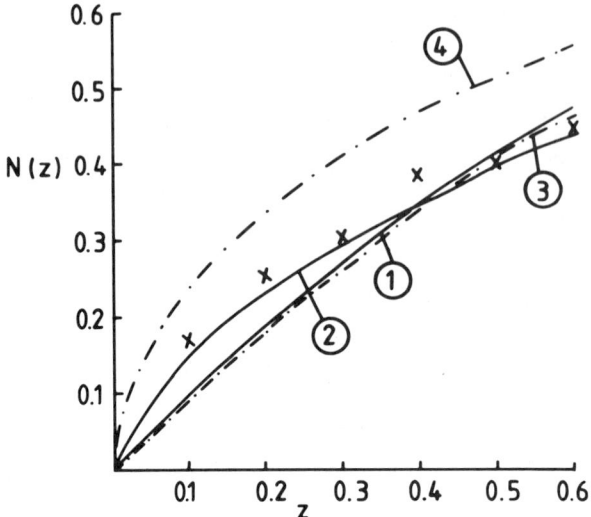

Figure 3. Comparison of experimental $N(z)$ points (marked by crosses) of anorthite with theoretical distributions ① and ② as calculated from (14) and (21) for space groups $P1$ and $P\bar{1}$ respectively. The dashed curves ③ and ④ are Sim's plots.

strong correlation exists between atomic positions in the unit cell. The theory may also be useful in putting relative intensities on to absolute scale in case of large-molecule structures.

Acknowledgement

One of the authors (GDN) wishes to express his thanks to Professor A.J.C. Wilson for the invitation to present this paper at a Microsymposium held during the XIII Congress and General Assembly of the International Union of Crystallography at Hamburg, West Germany. He is also thankful to Professor Dr H. Wondratschek for his encouragement and keen interest in the present work. The work was completed when he (GDN) was a Humboldt Fellow at the Institut für Kristallographie der Universität Karlsruhe, West Germany. It is a pleasure to thank Alexander von Humboldt-Stiftung for financial support.

References and Footnotes

1. Wilson, A.J.C. *Acta Cryst. 2*, 318-321 (1949)
2. Collin, R.L. *Acta Cryst. 8*, 499-502 (1955)
3. Sim, G.A. *Acta Cryst. 11*, 123-124 (1958)
4. Srinivasan, R. *Acta Cryst. 13*, 388-394 (1960)
5. French, S. and Wilson, K. *Acta Cryst. A34*, 517-525 (1975)
6. Wilson, A.J.C. *Acta Cryst. A37*, 808-810 (1981)
7. Wilson, A.J.C. *Acta Cryst. 17*, 1591-1592 (1964)
8. Nigam, G.D. *Indian J. Pure Appl. Phys. 10*, 655-656 (1972)

9. Nigam, G.D. and Wilson, A.J.C. *Acta Cryst. A36,* 832-833 (1980)
10. Jameson, M.B. and Penfold, B.R. *J. of the Chemical Society Part I,* 528-536 (1965)
11. Kempster, C.J.E., Megaw, H.D. and Radoslovich E. W. *Acta Cryst. 15,* 1005-1017 (1962)
12. Bloss, F.D. *Crystallography and Crystal Chemistry,* pp. 206-207, Holt, Rinehart and Winston, Inc. N.Y. (1971)

Appendix

$$I = \int \begin{matrix} \cos(2\pi s \cdot r) \\ \sin(2\pi s \cdot r) \end{matrix} \, dV. \qquad \text{(AI)}$$

Transforming it into spherical polar coordinates gives

$$I = \int_0^a \int_0^\pi \begin{matrix} \cos(2\pi sr \cos\psi) \\ \sin(2\pi sr \cos\psi) \end{matrix} \; 2\pi r^2 \sin\psi \, dr d\psi.$$

The lower integral is zero, since it is an odd function of ψ. Integrating first over ψ, we get

$$= \frac{2}{s} \int_0^a r \sin(2\pi sr) dr.$$

The above integral can be evaluated integrating by parts. Thus,

$$I = \frac{2}{s} \left[\frac{\sin(2\pi sa) - 2\pi sa \cos(2\pi sa)}{(2\pi s)^2} \right]. \qquad \text{(AII)}$$

With the help of (AII), the expressions for A and B can readily be derived.

Structure & Statistics in Crystallography, ISBN 0-940030-10-1,
Ed.. A. J. C. Wilson F. R. S., Adenine Press, ©Adenine Press 1985.

Centric, Bicentric and Partially Bicentric Intensity Statistics

Uri Shmueli
Department of Chemistry, Tel Aviv University
Ramat Aviv, 69 978 Tel Aviv, Israel

and

George H. Weiss
Physical Sciences Laboratory
Division of Computer Research and Technology
National Institutes of Health, Bethesda, MD 20205 USA

Abstract

The present status of generalized intensity statistics for centrosymmetric space groups is summarized by comparing probability density functions based on (i) expansions in terms of orthogonal polynomials, and (ii) characteristic functions of the structure factor and exact random-walk techniques. The latter approach is then applied to the derivation of an exact probability density function for the space group $P\bar{1}$, in the presence of one non-crystallographic center of symmetry in the asymmetric unit. This new bicentric distribution is further generalized to the important intermediate situation, in which only a part of the asymmetric unit is affected by non-crystallographic symmetry. The resulting partially bicentric distribution admits the centric and bicentric ones as special cases.

The theoretical probability functions, discussed and derived in this paper, are compared with distributions of the magnitude of the normalized structure factor that are recalculated from published structures or simulated by computer. The effects of the presence of outstandingly heavy atoms are emphasized throughout the comparisons of the various distributions with theoretical statistics.

Apart from graphical representations of the above comparisons, quantitative discrepancy criteria are also summarized. All these indicate that the performance of the new distributions is very good and that they seem to be ripe for application to problems arising in practical intensity statistics.

Introduction

The important role of structure-factor statistics, in the various stages of crystal-structure determination, is widely recognized. The probability density functions (hereafter: pdf's) of the structure factor and related quantities presently at the crystallographer's disposal range from asymptotic pdf's based on the central limit theorem (1-3), through orthogonal-polynomial expansions that take the asymptotic pdf's as their weight functions (4-8), to newly introduced pdf's that are based on Fourier and Fourier-Bessel expansions of the exact solution to the random walk problem (9-12).

All the recent developments in crystallographic statistics have aimed at an explicit introduction of the effects of space-group symmetry and atomic heterogeneity into the pdf's, to be used in comparisons with experimental data. When these general features are understood and allowed for, attention turns to the so called 'special' situations such as heavy atoms in special positions (13, 14) and effects of complete or partial non-crystallographic symmetry [for example, references (2) and (15)], all of which deserve much further study. Some of it is initiated in the present article.

The next section still relates to the general statistics, and specifically to the comparison of the centrosymmetric pdf's given by orthogonal-polynomial expansions with those based on random-walk techniques. In what follows, we present a short derivation of an exact pdf for the space group $P\bar{1}$, for the case of a single non-crystallographic center of symmetry and conclude the article with an extension of the above bicentric pdf to a more general one, which allows for local non-crystallographic centrosymmetry and accounts for the presence of additional atoms in the asymmetric unit of $P\bar{1}$ that do not possess any symmetry of internal arrangement. The new pdf's are presented as conveniently summable exact Fourier series.

Generalized distributions: no hypersymmetry

We consider here the status of generalized pdf's for $|E|$, where $|E|$ is the magnitude of the normalized structure factor, for centrosymmetric space groups. The pdf for $|E|$ has been generalized, for an arbitrary composition and space-group symmetry, by an orthogonal-polynomial expansion in which Wilson's (1) asymptotic centric pdf was taken as the weight function (6, 16). The general form of this generalized centrosymmetric pdf is

$$P(|E|) = \left(\frac{2}{\pi}\right)^{1/2} \exp\left(-\frac{|E|^2}{2}\right)\left[1 + \sum_{k=2}^{\infty} \frac{A_{2k}}{2^k(2k)!} H_{2k}\left(\frac{|E|}{\sqrt{2}}\right)\right] \qquad (1)$$

where $H_{2k}(x)$ are Hermite polynomials [see, *e.g.*, reference (17)] and the explicit dependence of the coefficients A_4, A_6, A_8 and A_{10} on the atomic composition of the asymmetric unit and the space-group symmetry has been published (6, 16, 18, 19). It should be pointed out that any number of additional coefficients can be constructed for triclinic, monoclinic and orthorhombic space groups (except *Fddd*) with the aid of known cumulant-moment relationships [see, *e.g.*, (20, (21)], since closed expressions for the required moments of the trigonometric structure factors are available (16). For all the remaining space groups, only A_4, A_6 and A_8 can be numerically evaluated from the available moments (18, 19). The status of availability of generalized noncentrosymmetric pdf's, of the above type, is closely analogous.

Centric & Bicentric Intensity Statistics

An alternative expression for centrosymmetric pdf's has recently been obtained by reintroducing random-walk models into crystallographic statistics (10, 12). The general form of this pdf, valid for any centrosymmetric space group, is

$$P(|E|) = a[1 + 2 \sum_{m=1}^{\infty} C_m \cos(\pi m a|E|)] \quad (2)$$

where the Fourier coefficient is based on the characteristic function for the structure factor F,

$$C(\omega_1) = \langle \exp(i\omega_1 F) \rangle, \quad (3)$$

which depends on the space-group symmetry and atomic composition [cf. (12)], and

$$a = \left(\sum_{j=1}^{N} f_j^2 \right)^{1/2} / \left(\sum_{j=1}^{N} f_j \right), \quad (4)$$

where f_j are the atomic scattering factors and N is the number of atoms in the unit cell. As shown elsewhere (10, 12), this approach has also been extended to non-centrosymmetric space groups.

Equations [1] and [2] are formally exact, but there are several important differences to be noted. Whereas all the terms of [2] have the same functional form, usually involving a product of Bessel functions, the complexity of the successive terms in [1] increases rapidly. For a given departure from the asympototic centric (1) distribution, and a desired accuracy, it is often easier to compute the required number of terms of equation [2] than to construct the required coefficients of [1]. However, the above mentioned complexity of the higher coefficients of [1] is the same for all the space groups while the characteristic functions [3] become increasingly difficult to evaluate as the number of terms in the trigonometric structure factor increases.

It therefore appears that, at the present stage of the derivations, the pdf's of the form [2] are preferable for space groups of low symmetries, and especially for structures with a small number of outstandingly heavy atoms; the approximate [1] may be inadequate in such situations [cf. (22) and (12)]. On the other hand, the approximate pdf [1] is the only one available for all the space groups, and it appears to be sufficiently accurate for those of higher symmetry: the departures from the asymptotic pdf's [reference (1)] are then much smaller, for the same degree of atomic heterogeneity (23, 6).

The performance of the pdf [2], for a highly heterogeneous $P\bar{1}$ structure [cf. (10)], is shown in Figure 1 (a) and that of the pdf [1], truncated to the first five terms, is illustrated for a significantly more heterogeneous monoclinic structure, in Figure 1(b). The first part of Table I summarizes some quantitative discrepancy criteria, the calculated values of χ^2 and R (10), for these comparisons.

The results given in Figure 1(*a*), as well as other examples [cf. (10)], show that the introduction of the random-walk Fourier pdf's to intensity statistics has greatly advanced the feasibility of resolving $P\bar{1}$ *vs* $P1$ ambiguities, which are well known to be strongly affected by atomic heterogeneity. Although the approximate pdf's based on the first few terms of equation [1] also provide qualitatively correct indications,

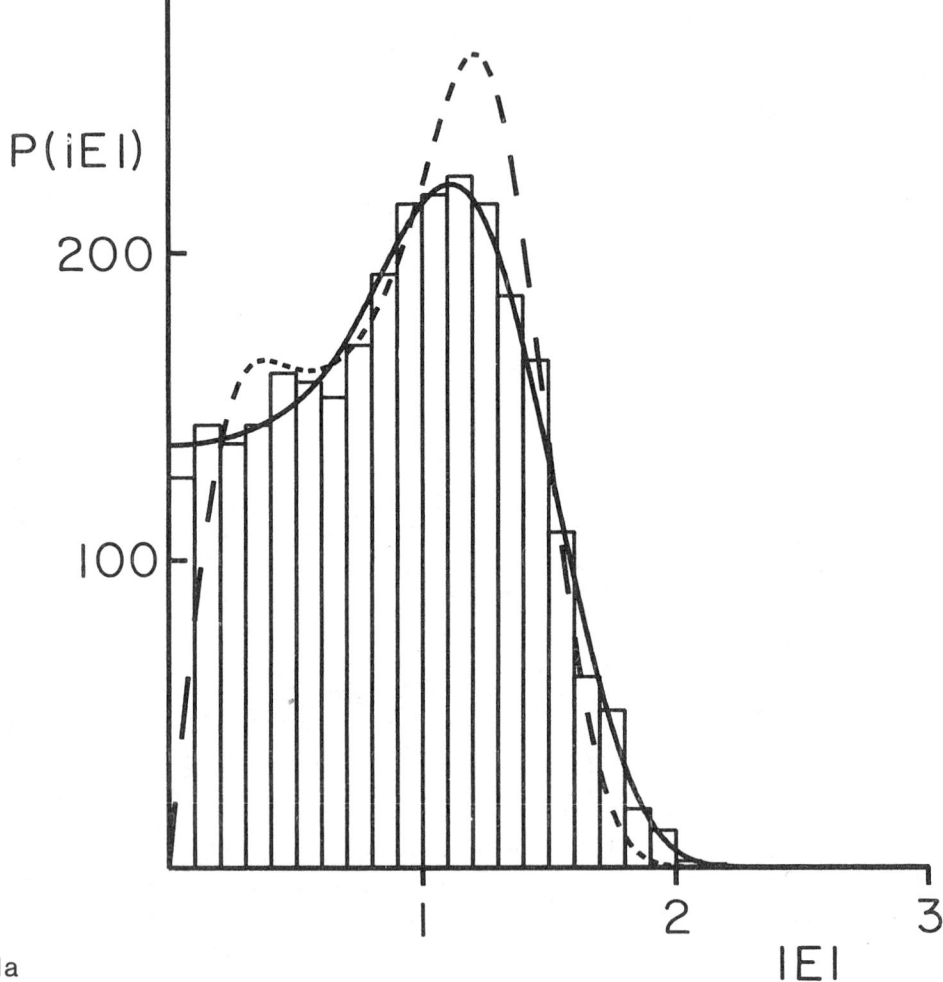

Figure 1. *Comparison of histograms of recalculated $|E|$ values with theoretical distributions.* The histograms were produced by a modified version (24) of program NORMAL of the MULTAN80 system, from structure factors that were recalculated from published atomic parameters of the structures referred to below. The height of each rectangle equals the number of $|E|$ values that lie in the corresponding histogram channel. (*a*) Recalculated data for $C_6Cl_2N_4O_4$ platinum, space group $P\bar{1}$ with $Z = 2$ [reference (29)], compared with Fourier (solid, $P\bar{1}$) and Fourier-Bessel (dashed, $P\bar{1}$) random-walk pdf's (10). (*b*) Recalculated data for $[NH_4][SO_4]_2 \cdot 4H_2O$ uranium, space group $P2_1/c$ with $Z = 4$ [reference (30)], compared with five-term Hermite (solid, $P2_1/c$) and Laguerre (dashed, Pc) generalized pdf's (16). All the theoretical distributions are scaled to the corresponding histograms.

Centric & Bicentric Intensity Statistics

in the triclinic case, their performance is definitely inferior to that of the Fourier pdf's (16, 12).

The departures from the asymptotic statistics (1) for the monoclinic space groups may also be significantly large [see, *e.g.,* reference (6)] but, as shown by Fig. 1(*b*) and Table I for a uranium compound, they may already be accounted for by the first five terms of the Hermite pdf [1]. Similar results were obtained for several monoclinic compounds, of comparable degrees of heterogeneity, although the discrepancy criteria tend to be worse than for random-walk pdf's so far examined, and regions of slightly negative probability are still present [*cf.* Fig. 1(*b*)]. A replacement of Hermite-Laguerre pdf's (6, 16) by Fourier and Fourier-Bessel random-walk

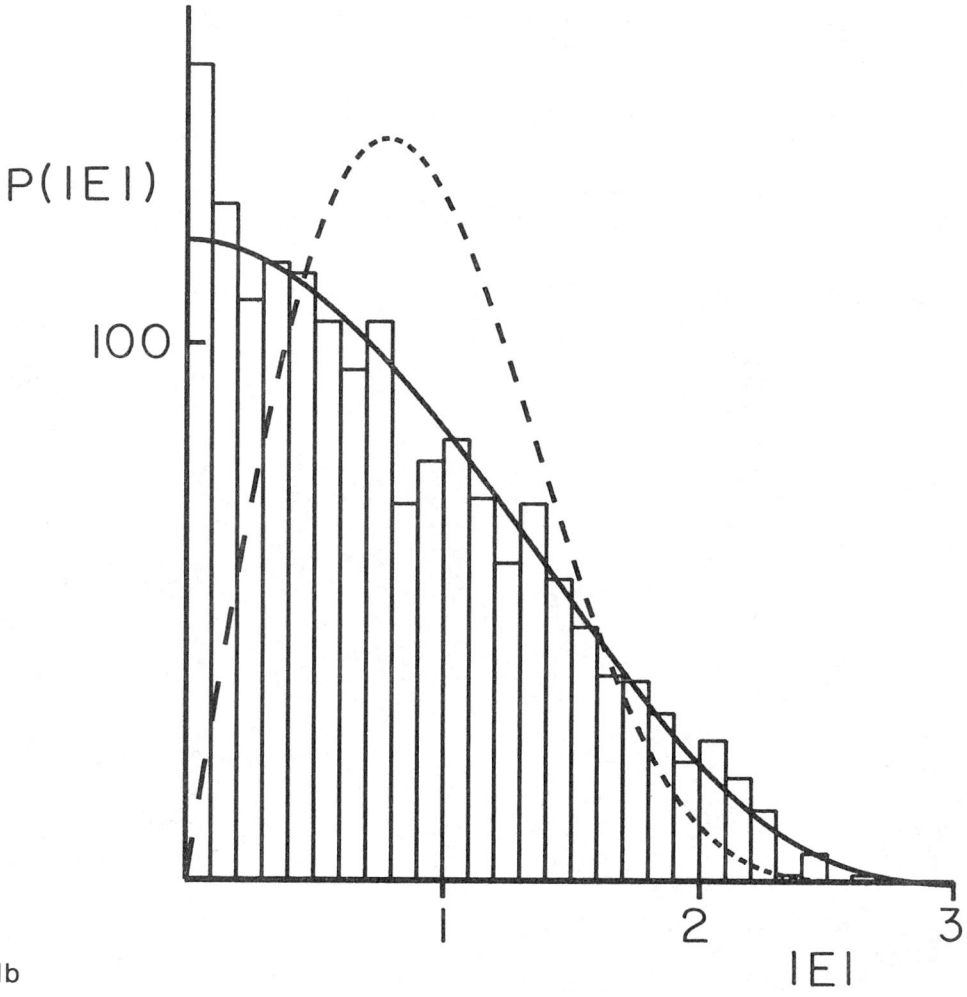

Figure 1 continued from previous page.

Table I

Quantitative discrepancy criteria for the comparison of the recalculated and simulated histograms with the theoretical pdf's. The criteria listed are χ^2, k—effective number of channels used in the calculation of χ^2 and the R factor [cf. references (10) and (25)]. It is recalled that the expected value of is $k - 1$, and the '2σ' limit of χ^2 is $k - 1 + 2[2(k - 1)]^{1/2}$ (see above references). The comparison for a given histogram channel is made between the height of the corresponding rectangle and the value(s) of the scaled theoretical pdf(s), taken at the midpoint of the channel. The Table refers directly to the Figures.

(A) Recalculated distributions

Figure 1(a)

Histogram—solid ($P\bar{1}$):	$\chi^2 = 9.6$	$k = 20$	$R = 0.047$
Histogram—dashed ($P1$):	$\chi^2 = 415.0$	$k = 19$	$R = 0.183$

Figure 1(b)

Histogram—solid ($P2_1/c$):	$\chi^2 = 32.3$	$k = 25$	$R = 0.122$
Histogram—dashed (Pc):	$\chi^2 = 1836.0$	$k = 22$	$R = 0.542$

(B) Bicentric distributions

Figure 2

Histogram—dashed[1]	$\chi^2 = 27.1$	$k = 26$	$R = 0.086$
Histogram—solid[2]	$\chi^2 = 381.7$	$k = 30$	$R = 0.274$

(C) Partially bicentric distributions

Figure 3(a)

Histogram—dashed[3]	$\chi^2 = 24.7$	$k = 28$	$R = 0.080$
Histogram—solid[2]	$\chi^2 = 84.4$	$k = 30$	$R = 0.110$

Figure 3(b)

Histogram—dashed[3]	$\chi^2 = 21.7$	$k = 30$	$R = 0.057$
Histogram—solid[AB]	$\chi^2 = 103.8$	$k = 30$	$R = 0.223$
Histogram—solid[AC]	$\chi^2 = 141.4$	$k = 29$	$R = 0.183$

1—random-walk Fourier bicentric pdf.
2—asymptotic bicentric pdf [reference (2)].
3—random-walk Fourier partially bicentric pdf.
AB—as 2.
AC—asymptotic centric pdf [reference (1)].

pdf's (12), for all the space groups of low symmetry, is being considered in current work on application software related to intensity statistics (24).

It must be noted that both Figures, and the first part of Table I, are based on data recalculated from published structures and the comparisons are probably not seriously affected by the (admitted) omission of unobserved reflections from the data, and by effects of statistical fluctuations on the measured intensities (25). There is no apparent non-crystallographic symmetry in the structures that underly Figures 1(a) and 1(b), and all the atoms are located in general Wyckoff positions—as required by the available versions of the above pdf's.

Centric & Bicentric Intensity Statistics

The bicentric distribution

An important factor, not accounted for by the pdf's given above, is the presence of some well defined non-crystallographic symmetry within the asymmetric unit of the centrosymmetric crystal. The most extensively studied example of such additional symmetry, in the context of intensity statistics, is the presence of one or more non-crystallographic centers of symmetry. The resulting asymptotic probability density functions were given by Lipson and Woolfson (26) and by Rogers and Wilson (2) for the presence of one non-crystallographic center of symmetry, and were generalized by the latter authors (2) for the presence of several non-crystallographic centers in the asymmetric unit of $P\bar{1}$. The subject was recently reinvestigated (27) and the previously proposed asymptotic bicentric distribution (2) was extended to a series involving orthogonal polynomials, with coefficients that take into account the atomic composition and space-group symmetry. The latter development is conceptually similar to those underlying equation [1], and hence the derivation of an exact bicentric pdf for the space group $P\bar{1}$ is of interest for its own sake, as well as for reasons connected with the particular sensitivity of this space group to atomic heterogeneity.

Such a derivation, starting from the above mentioned characteristic function approach and using the method of Fourier representations of crystallographic pdf's (12), turned out to be rather straightforward. We assume that (i) the asymmetric unit of $P\bar{1}$ consists of two identical subunits that are related by a non-crystallographic center at, say, d, and (ii) all the atoms, as well as the additional center, occupy positions that permit to treat the fractional parts of the scalar products $h \cdot r_j$ where h is the diffraction vector, as random variables uniformly distributed in the [0,1] range; this is, in practice, equivalent to the assumption that only general Wyckoff positions are involved [*cf.* reference (10)].

The structure factor for the above arrangement can most conveniently be written as

$$F(h) = 2\sum_{j=1}^{N/4} f_j[\cos 2\pi h \cdot r_j + \cos 2\pi h \cdot (2d - r_j)]$$

$$= 4\sum_{j=1}^{N/4} f_j \cos(2\pi h \cdot d)\cos[2\pi h \cdot (r_j - d)] ,$$

(5)

where N is the number of atoms in the unit cell, and f_j and r_j are atomic scattering factors and position vectors respectively (11). If we further define $\varphi = 2\pi h \cdot d$ and $\theta_j = 2\pi h \cdot r_j$ as the random variables of the problem, the characteristic function can be written as

$$C(\omega_1) = \langle \exp[4i\omega_1 \sum_{j=1}^{N/4} f_j \cos\varphi \cos(\theta_j - \varphi)] \rangle$$

(6)

and integration over the variables $\theta_1, \ldots, \theta_{N/4}$ leads to:

$$C(\omega_1) = \frac{2}{\pi} \int_0^{\pi/2} [\prod_{j=1}^{N/4} J_o(4\omega_1 f_j \cos\varphi)] d\varphi \qquad (7)$$

where $J_o(x)$ is a Bessel function. Making use of the fact that the possible non-zero values of F are bounded in the $[-S_1, S_1]$ range, where $S_1 = \sum_{j=1}^{N} f_j$, we can expand the pdf of F, as detailed elsewhere [see reference (12)], in a Fourier series. The series has the same form as that given in equation [2] and the Fourier coefficients for the present problem are given by

$$C_m = \frac{2}{\pi} \int_0^{\pi/2} [\prod_{j=1}^{N/4} J_o(\frac{4\pi m f_j}{S_1} \cos\varphi)] d\varphi . \qquad (8)$$

Some further details of the derivation are given in reference (11), where it is also shown that equation [2] with coefficients [8] reduces to the asymptotic bicentric pdf given in (2), provided it is assumed that all the atoms are of the same kind and the number of atoms in the unit cell is large.

It will be shown below that if the asymmetric unit of $P\overline{1}$ contains just two equal atoms, of necessity related by a non-crystallographic center, the resulting bicentric pdf reduces to a centric one.

The departure of P(|E|), given by equation [1], from the Wilson (1949) centric distribution (1) increases with increasing atom heterogeneity (23, 10) and the same is true, albeit to a somewhat smaller extent, when the above derived bicentric pdf is compared with the asymptotic Rogers-Wilson bicentric distribution (2). Figure 2 shows a comparison of these two theoretical distributions with a computer-simulated bicentric distribution, based on an asymmetric subunit containing fourteen carbons and one uranium atom. Details of the numerical integration of (8) and the simulation procedure are given in reference (11).

It is evident that the performance of the exact bicentric pdf is very good, even when quantitative discrepancy measures χ^2 and R (10, 25) are considered (see Table I). An extension of the above procedure to the presence of several non-crystallographic centers (2) is being investigated by the present authors.

Intermediate or partially bicentric distributions

The probability measures discussed in the previous sections are still idealized, at least in some cases, as they do not account for partial hypersymmetry that so often characterizes real structures. As is well known, centrosymmetric asymmetric units may range from assemblies of atoms with no internal symmetry (as is most usually

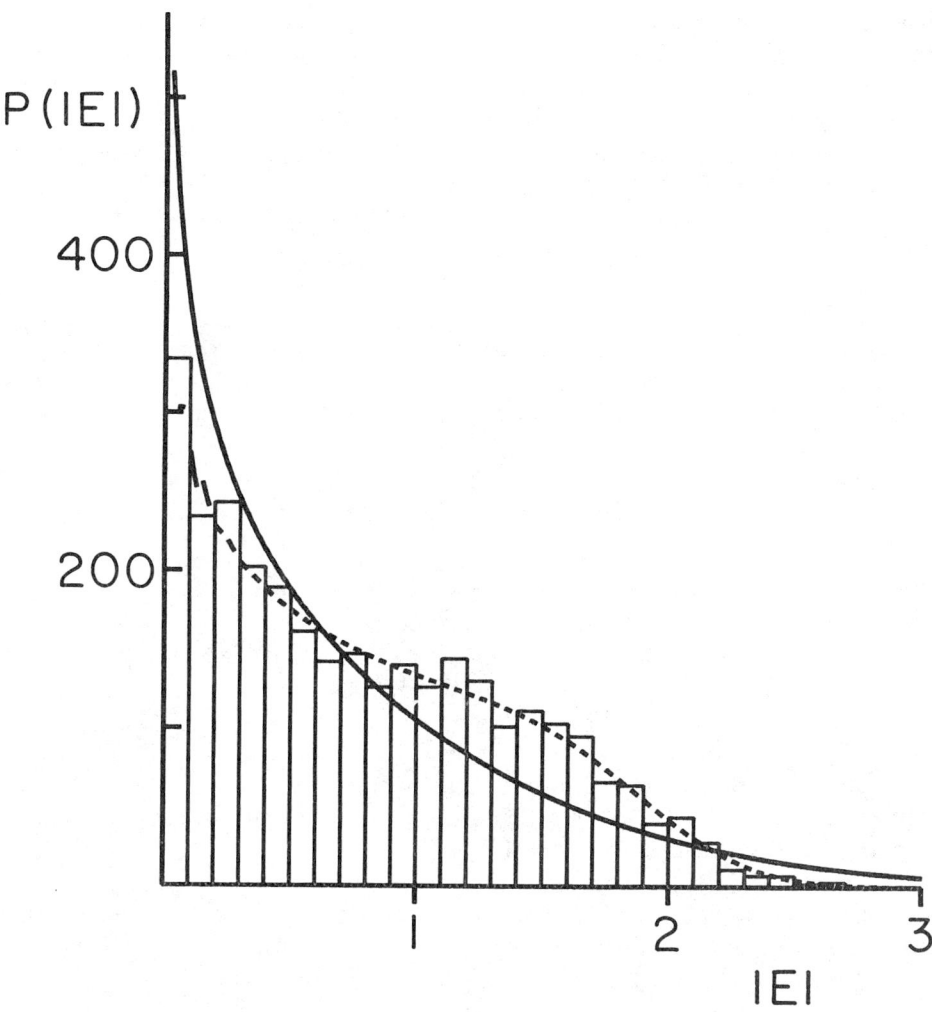

Figure 2. *Simulated and theoretical bicentric distributions of* $|E|$. The simulation is based on equation [5] and on subsequent replacements of the scalar products $\mathbf{h}\cdot\mathbf{r}_j$ and $\mathbf{h}\cdot\mathbf{d}$ by random numbers. The histogram is construed from 3000 simulated $|E|$ values. The solid curve corresponds to the asymptotic bicentric pdf (2) and the dashed one is the Fourier random-walk bicentric pdf (equ. [2]) with coefficients given by [8]. Both theoretical distributions are scaled to the histogram.

assumed in practical statistical tests), through units containing important, or dominant, centrosymmetric fragments, to purely bicentric arrangements (*e.g.*, a centrosymmetric molecule with the molecular center in a general position). An intermediate, or partially hypersymmetric, arrangement may of course be caused by any element of non-crystallographic symmetry but only the simplest case of a partially bicentric arrangement will be considered in what follows.

The derivation of the partially bicentric pdf is a direct extension of the derivation of the bicentric pdf, given in the previous section [see also references (11) and (12)].

Let the unit cell of $P\bar{1}$ contain $N = N_B + N_R$ atoms, where N_B of them constitute two centrosymmetric fragments, containing non-crystallographic centers of symmetry and the remaining N_R atoms are related in pairs by the crystallographic center of symmetry. All the atoms, as well as the non-crystallographic center at $\boldsymbol{d}$, are again assumed to occupy general Wyckoff positions. Using [5], the structure factor for the above arrangement can be written as

$$F(h) = 4 \sum_{j=1}^{N_B/4} f_j \cos(2\pi \boldsymbol{h}\cdot\boldsymbol{d})\cos[2\pi \boldsymbol{h}\cdot(\boldsymbol{r}_j - \boldsymbol{d})] + 2 \sum_{k=1}^{N_R/4} f_k \cos 2\pi \boldsymbol{h}\cdot\boldsymbol{r}_k . \qquad (9)$$

Since the fractional parts of the scalar products $\boldsymbol{h}\cdot\boldsymbol{d}$, $\boldsymbol{h}\cdot\boldsymbol{r}_j$, and $\boldsymbol{h}\cdot\boldsymbol{r}_k$ can now be taken as uniform in the [0,1] range, and introducing the notation: $\varphi = 2\pi \boldsymbol{h}\cdot\boldsymbol{d}$, $\theta_j = 2\pi \boldsymbol{h}\cdot\boldsymbol{r}_j$ and $\psi_k = 2\pi \boldsymbol{h}\cdot\boldsymbol{r}_k$, the characteristic function for the present problem is given by

$$C(\omega_1) = \langle \exp i\omega_1 [4 \sum_{j=1}^{N_B/4} f_j \cos\varphi \cos(\theta_j - \varphi) + 2 \sum_{k=1}^{N_R/2} f_k \cos\psi_k] \rangle , \qquad (10)$$

and evaluation of the readily soluble definite integrals leads to

$$C(\omega_1) = \Big\{ \prod_{k=1}^{N_R/2} J_o(2\omega_1 f_j) \Big\} \Big\{ \frac{2}{\pi} \int_0^{\pi/2} [\prod_{j=1}^{N_B/4} J_o(4\omega_1 f_j \cos\varphi)] d\varphi \Big\}. \qquad (11)$$

The resulting characteristic function is thus simply the product of centric and bicentric functions, each involving the number of atoms that has been defined in our model [see also reference (12), in this book].

Following the same arguments that are used in the previous section, the Fourier series for the pdf of $|E|$ has the same functional form as that in equation [2], and the Fourier coefficients for the partially bicentric distribution are given by

$$C_m = \Big\{ \prod_{k=1}^{N_R/2} J_o(\frac{2\pi m f_j}{S_1}) \Big\} \Big\{ \frac{2}{\pi} \int_0^{\pi/2} [\prod_{j=1}^{N_B/4} J_o(\frac{4\pi m f_j}{S_1} \cos\varphi)] d\varphi \Big\}. \qquad (12)$$

It follows that not only the derivation but also the numerical evaluation of the partially bicentric pdf is only marginally more complicated than that of the pure bicentric pdf for $P\bar{1}$.

An interesting special case of an apparently partially bicentric distribution is that arising from the presence of two outstandingly heavy identical atoms and a number of light ones, in the asymmetric unit. This is typical for many organometallic compounds that appear in the recent literature. Since the pair of heavy atoms is of necessity related by a non-crystallographic center of symmetry, this pair might be regarded as a 'bicentric' fragment of the asymmetric unit. In the case of the space group $P\bar{1}$, there are four such atoms in the unit cell and the 'bicentric' factor of the Fourier coefficient [12] is given by

$$C_{Bm} = \frac{2}{\pi}\int_0^{\pi/2} J_o\left(\frac{4\pi m f_H}{S_1}\cos\varphi\right)d\varphi, \quad (13)$$

where f_H is the scattering factor of a heavy atom. Since only a single Bessel function has remained in the integrand, use can be made of the known definite integral (28)

$$\int_0^{\pi/2} \cos(2\mu x)J_{2\nu}(2a\cos x)dx = \frac{\pi}{2}J_{\nu+\mu}(a)J_{\nu-\mu}(a), \quad \text{Re}(\nu) > -\frac{1}{2} \quad (14)$$

to obtain the required 'bicentric' factor:

$$C_{Bm} = J_o^2\left(\frac{2\pi m f_H}{S_1}\right). \quad (15)$$

The result is simply a contribution of two atoms with scattering factors f_H to the Fourier coefficient of the pdf for $P\bar{1}$ in the absence of hypersymmetry [cf. reference (10)], i.e. to the first product in (12). In other words, asymmetric units of $P\bar{1}$ that contain two equal and outstandingly heavy atoms do not require any 'bicentric' treatment, no matter how heavy these atoms are, provided there are no additional atoms that follow the bicentric arrangement. It also follows that the minimum size of a bicentric fragment of the asymmetric unit of $P\bar{1}$ is two pairs of atoms related by a non-crystallographic center of symmetry, and the same is of course true if all the asymmetric unit of $P\bar{1}$ consists of a bicentric fragment (see above).

Several distributions that arise from such partially bicentric arrangements have been simulated, and two of them are compared in Figure 3 with the partially bicentric Fourier pdf derived above, and with the Rogers-Wilson bicentric pdf (2). The first example [Fig. 3(a)] deals with an asymmetric unit of $P\bar{1}$ that contains four equal heavy atoms, which form a centrosymmetric group, and fifteen light atoms outside the above 'bicentric' quartet. The ratio of the atomic numbers, Z(heavy)/Z(light) was taken as 15 and one third (i.e. the composition approximates U_4C_{15}). Figure 3(a) shows that the simulated distribution arising from the above partially bicentric arrangement agrees very well with the Fourier pdf [2], with coefficients given by [12]. The distribution is, however, also remarkably close to the Rogers-Wilson pdf (2). This is so, since the model consists, to a very rough approximation, of four equal atoms in a bicentric arrangement, and hence the distribution should approximately follow the asymptotic bicentric pdf (2). As the Z(heavy)/Z(light) ratio decreases, for the same arrangement, the discrepancy between the partially bicentric Fourier and the asymptotic bicentric pdf's increases, until in the equal-atom limit the simulated distribution becomes expectedly close to the asymptotic centric pdf (1).

The second example concerns a hypothetical equal-atom $P\bar{1}$ structure, with an asymmetric unit that consists of $N_B/2$ atoms forming a centrosymmetric fragment and $N_R/2$ atoms not related by any non-crystallographic symmetry. Such situations occur rather frequently in molecular crystals of organic compounds. A series of

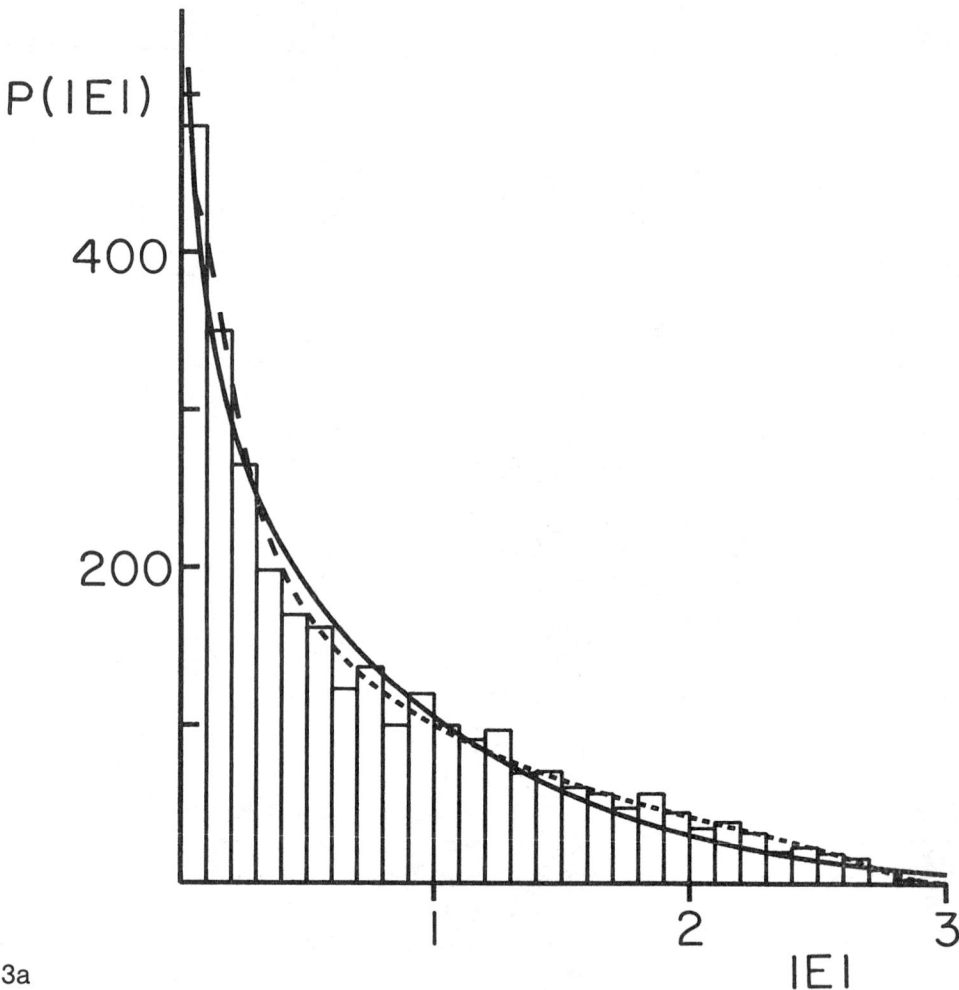

3a

Figure 3. *Simulated and theoretical partially bicentric distributions of* $|E|$. The simulations are based on equation [9] and on subsequent replacements of the scalar products, appearing in [9], by random numbers. Each histogram is constructed from 3000 simulated $|E|$ values. The solid curves correspond to the asymptotic bicentric [reference (2)] and centric [reference (1)] pdf's, and the dashed ones represent the Fourier random-walk partially bicentric pdf's [2] with coefficients given by [12]. The theoretical distributions are scaled to the corresponding histograms. (a) Asymmetric unit of $P\bar{1}$: four heavy atoms in a centrosymmetric fragment, and fifteen light unrelated one ($N_B = 8$, $N_R = 30$, Z(heavy)/Z(light) = 15 1/3). The asymptotic centric pdf is not given in this Figure. (b) Asymmetric unit of $P\bar{1}$: twenty six atoms in a centrosymmetric fragment and four unrelated one ($N_B = 52$, $N_R = 8$, all the atoms are the same kind). The letters AB and AC denote the asymptotic bicentric and centric pdf's respectively.

simulations has been done for a $P\bar{1}$ asymmetric unit containing 30 equal atoms, while varying the relative proportions of the 'bicentric' and 'centric' components. These simulations indicate that an appreciable deviation of the histogram of simulated $|E|$ values from the Wilson asymptotic centric pdf (1) appears when the size of the centrosymmetric fragment reaches 20 atoms, and the departure from the asymptotic bicentric pdf (2) then begins to decrease. All these histograms are very well accounted for by the partially bicentric distribution given by equations [2] and [12]. Figure 3(b) shows the result of such a calculation for an asymmetric unit containing a 26-atom centrosymmetric fragment and four atoms not related by any symmetry. The discrepancies between the simulated histogram and the asymptotic centric and bicentric pdf's are both significant and comparable, while the agreement with the partially bicentric pdf is very good.

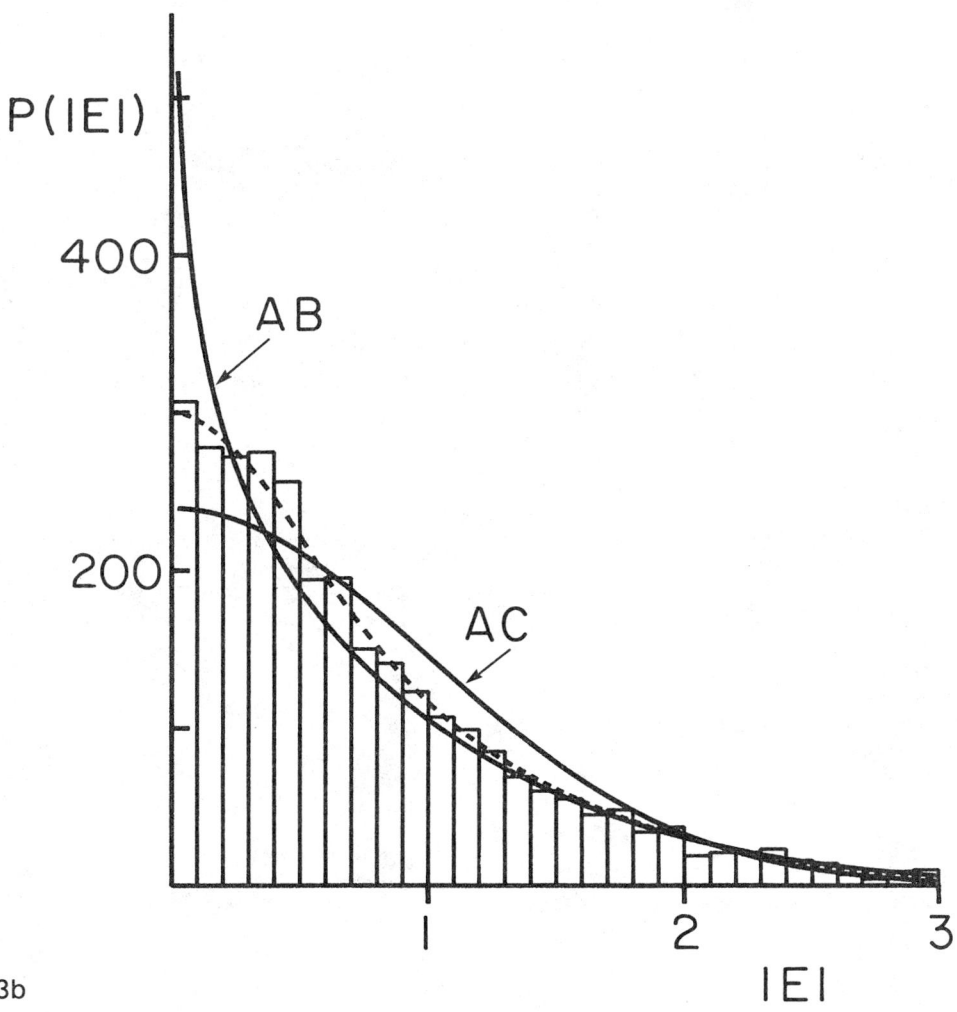

Figure 3 continued from previous page.

The partially bicentric distribution, derived and illustrated in this section, naturally admits the exact centric and bicentric distributions as special cases and its implementation in crystallographic application software may be helpful in explaining some 'mysterious' intensity statistics that are not unfrequently encountered during the process of crystal structure determination.

References and Footnotes

1. Wilson, A.J.C., *Acta Cryst. 2,* 318-321 (1949).
2. Rogers, D. & Wilson, A.J.C., *Acta Cryst. 6,* 439-449 (1953).
3. Wilson, A.J.C., *Acta Cryst. A36,* 945-946 (1980).
4. Hauptman, H. & Karle, J., *Solution of the Phase Problem. I. The Centrosymmetric Crystal,* ACA Monograph No. 3, Polycrystal Book Service, Pittsburgh (1953).
5. Klug, A., *Acta Cryst. 11,* 515-543, (1958).
6. Shmueli, U. & Wilson, A.J.C., *Acta Cryst. A37,* 342-353, (1981).
7. Shmueli, U. & Wilson, A.J.C. In *Conformation in Biology,* Ed. R. Srinivasan and R.H. Sarma, Adenine Press, New York, pp. 383-388 (1982).
8. Shmueli, U. & Wilson, A.J.C., *Acta Cryst. A39,* 225-233 (1983).
9. Weiss, G.H. & Kiefer, J.E., *J. Phys. A 16,* 489-495 (1983).
10. Shmueli, U., Weiss, G.H., Kiefer, J.E. & Wilson, A.J.C., *Acta Cryst. A40,* 551-560 (1984).
11. Shmueli, U., Weiss, G.H. & Kiefer, J.E., *Acta Cryst. A* 41, 55-59 (1985).
12. Weiss, G.H., Shmueli, U., Kiefer, J.E. & Wilson, A.J.C. (1985). This volume, pp. 23-42.
13. Sim, G.A., *Acta Cryst. 11,* 123-124 (1958).
14. Foster, F. & Hargreaves, A., *Acta Cryst. 16,* 1124-1133 (1963).
15. Srinivasan, R. & Parthasarathy, S., *Some Statistical Applications in X-Ray Crystallography,* Pergamon Press, Oxford, p. 57 (1976).
16. Shmueli, U., *Acta Cryst. A38,* 362-371 (1982).
17. Abramowitz, M. & Stegun, I., *Handbook of Mathematical Functions,* Dover, New York (1972).
18. Shmueli, U. & Kaldor, U., *Acta Cryst. A37,* 76-80 (1981).
19. Shmueli, U. & Kaldor, U., *Acta Cryst. A39,* 619-621 (1983).
20. Kendall, M.G. & Stuart, A., *The Advanced Theory of Statistics,* Vol. 1, 3rd ed., Charles Griffin, London (1969).
21. Shmueli, U. & Wilson, A.J.C., In *Crystallographic Statistics: Progress and Problems,* Ed. S. Ramaseshan, M.F. Richardson and A.J.C. Wilson, Indian Academy of Sciences, Bangalore, pp. 83-97 (1982).
22. Shmueli, U., In *Crystallographic Statistics: Progress and Problems,* Ed. S. Ramaseshan, M.F. Richardson and A.J.C. Wilson, Indian Academy of Sciences, Bangalore, pp. 53-82 (1982).
23. Shmueli, U., *Acta Cryst. A35,* 282-286 (1979).
24. Shmueli, U. In preparation. (1984).
25. Wilson, A.J.C. (1985). This volume, pp. 87-94.
26. Lipson, H. & Woolfson, M.M., *Acta Cryst. 5,* 680-682 (1952).
27. Ghosh, S. & Nigam, G.D., *Zeitschrift für Kristallographie 163,* 61-68 (1983).
28. Gradshteyn, I.S. & Ryzhik, I.M., *Tables of Integrals, Series and Products,* Academic Press, New York, entry: 6.681(1) (1980).
29. Faggiani, R., Lippert, B. & Lock, C.J.L., *Inorg. Chem. 19,* 295-300 (1980).
30. Bullock, J.I., Ladd, M.F.C., Povey, D.C. & Storey, A.E., *Inorganica Chimica Acta 43,* 101-108 (1980).

Structure & Statistics in Crystallography, ISBN 0-940030-10-1,
Ed., A. J. C. Wilson F. R. S., Adenine Presss, ©Adenine Press 1985.

Non-crystallographic Translational Symmetry: Effects on Diffraction-Intensity Statistics

G. Cascarano, G. Giacovazzo & M. Luić (1)

Dipartimento Geomineralogico, Università di Bari,
70121 Bari, Italy

Abstract

Non-crystallographic translational symmetry may or may not give rise to a substructural unit cell. A mathematical model is described which is able to deal with both situations and provide statistical relations for diffraction intensities which, in favourable cases, uniquely characterize the translational symmetry.

Introduction

Crystal structures having superstructure characteristics were divided by Buerger (2,3) into two parts: (i) the substructure, comprising that part of the electron density which conforms to the periodicity of the sub-cell, and (ii) the complement structure comprising the rest of the electron density. Accordingly, the reciprocal lattice {H} may be divided into two subsets: (i) the subset {A} belonging to the reciprocal superlattice of the substructure and (ii) the set {B} ("excess points" in Buerger's notation). The Fourier transform of the electron-density distribution of the substructure is zero everywhere except at the set {A}, while the Fourier transform of the complement structure contributes to the entire set {H}. The following relation holds:

$$\rho_{sub}(r) \leq \rho(r)$$

where $\rho(r)$ is the electron-density distribution of the full structure.

A general algebra of the substructures was described by Taxer (4), who introduced the concept of ideal substructure. According to Taxer, the ideal substructure is defined as a substructure with a higher space-group symmetry than that of the full structure and/or with a smaller unit cell being the subcell. Therefore the space group of the full structure is a subgroup of that of the substructure.

Practical substructures do not always exactly fit the above definition, both because the symmetrical arrangement of the atoms is disturbed by small deformations and because of substitution of some atoms by other kind of atoms. In this view Taxer introduced the concept of idealized substructure which complies with Buerger-Taxer definitions provided suitable correction functions have applied to the actual substructure in order to shift slightly misplaced atoms and correct electron density. The above considerations proved very useful in describing, comparing and solving a variety of crystal structures.

Translational symmetry additional to that required by space-group symmetry does not always produce substructure effects. Indeed crystallographic situations occur where the presence of non-crystallographic translational symmetry does not allow the identification of any, even non-ideal, substructure. These cases are of great interest in crystallography, both from a statistical point of view [the structure-factor intensities are systematically affected by structure regularities (5,6)], and from the direct-methods point of view (the *a-priori* identification of a structure regularity can make easier the crystal-structure solution).

This paper aims at: (i) describing systematic effects of non-crystallographic translations on the structure-factor intensities and (ii) extracting information on non-crystallographic translations by analysis of the structure-factor intensities.

The Mathematical Model

Let us suppose that a non-negligible amount of electron density $\rho(r)$, say $\rho_p(r)$, satisfies the condition

$$\rho_p(r) \simeq \rho_p(r+u) .$$

Then we say that a pseudotranslation u occurs, given by

$$u = \frac{v'_1}{\mu'_1} a + \frac{v'_2}{\mu'_2} b + \frac{v'_3}{\mu'_3} c , \qquad (1)$$

where a, b, c are the vectors of the unit cell, v'_i and μ'_i are integer numbers whose greatest common divisor is unity, and

$$0 \leq v'_i < \mu'_i \qquad i = 1, 2, 3.$$

The index n of the pseudotranslation is the smallest integer for which $n u$ is a lattice vector: it coincides with the least common multiple of μ'_1, μ'_2, μ'_3. In $P1$ n gives the number of sites related by u to a given positional vector r. Let us now suppose that in a space group of order m more than one independent pseudotranslations are simultaneously present (two pseudotranslations are independent of one another if the presence of the first is not the necessary consequence of the presence of the second and *vice-versa*). u_i will denote the ith independent pseudotranslation and n_i its order. Let p be the overall number of atoms, symmetry-equivalent included, whose positions are related by the pseudotranslations u_i, $i=1,2,...$, and let t_p be the number of independent atoms which generate the p atoms when the pseudotrans-

Structure Factors & Pseudotranslations

lation u_i, $i=1,2,...$ and the symmetry operators C_s, $s=1,...,m$ are applied. Then each r_j of the t_p atoms will generate $n_1 \cdot n_2 \cdot n_3 \cdot ...$ equivalent atoms in positions

$$(r_j + \nu_1 u_1 + \nu_2 u_2 + \nu_3 u_3 + ...), 0 < \nu_i \le n_i - 1$$

plus symmetry equivalent atoms (because of the space group symmetry). In all, $mn_1 n_2 n_3...$ positions are generated given by

$$C_s(r_j + \nu_1 u_1 + \nu_2 u_2 + \nu_3 u_3 + ...), 0 < \nu_i \le n_i - 1 \\ s=1,2,...,m.$$

$C_s \equiv (R_s, T_s)$ is the sth symmetry operator: R_s is its rotational part, T_s the translational part. For the sake of simplicity we have supposed that no atom lies on a special crystallographic site.

In accordance with a recent paper (7) the structure-factor equation may be written in the form

$$F_h = \sum_{j=1}^{t_p + t_q} f_j g_j \qquad (2)$$

where

$$g_j = \sum_{s=1}^{m} \frac{\sin n_1 \pi h R_s u_1}{\sin \pi h R_s u_1} \cdot \frac{\sin n_2 \pi h R_s u_2}{\sin \pi h R_s u_2} \cdot \frac{\sin n_3 \pi h R_s u_3}{\sin \pi h R_s u_3} \cdots$$

$$\times \exp 2\pi i h C_s \left(r_j + \frac{n_1 - 1}{2} u_1 + \frac{n_2 - 1}{2} u_2 + \frac{n_3 - 1}{2} u_3 + ... \right)$$

for $j \le t_p$,

$$g_j = \sum_{s=1}^{m} \exp 2\pi i h C_s r_j$$

for $j > t_p$.

In the above notation q is the number of atoms, symmetry-equivalent included, whose positions are not related by any pseudotranslation and t_q is the number of independent atoms which generate the q atoms by application of the symmetry operators C_s. The expression (2) emphasizes the fact that the number of the primitive tridimensional random variables in our approach is $t_p + t_q$.

The expected mean value of the structure-factor intensity $|F_h|^2$ is given by

$$\langle |F_h|^2 \rangle = \sum_{j=1}^{t_p+t_q} f_j^2 \langle |g_j|^2 \rangle = \epsilon_h (\Sigma_{t_p} \cdot \delta + \Sigma_q) \qquad (3)$$

where

ϵ_h is the statistical weight of the reflexion h,

$$\delta = \sum_{s=1}^{m} \frac{\sin^2 n_1 \pi hR_s u_1}{\sin^2 \pi hR_s u_1} \cdot \frac{\sin^2 n_2 \pi hR_s u_2}{\sin^2 \pi hR_s u_2} \cdot \frac{\sin^2 n_3 \pi hR_s u_3}{\sin^2 \pi hR_s u_3} \cdots$$

$$\Sigma_{t_p} = \sum_{j=1}^{t_p} f_j^2, \ \Sigma_q = \sum_{j=1}^{q} f_j^2.$$

The expression (3) may be written down in a more useful form:

$$\langle |F_h|^2 \rangle = \epsilon_h (\alpha_h \Sigma_p + \Sigma_q) \qquad (4)$$

where

$$\alpha_h = \frac{(n_1 n_2 n_3 \ldots)}{m} \gamma_h \qquad (5)$$

and γ_h is the number of times for which the algebraic congruences

$$hR_s u_i \equiv 0 (\text{mod } 1) \text{ for } i=1,2,3,\ldots \qquad (6)$$

are simultaneously satisfied when s varies from 1 to m. The maximum value of γ is m: consequently the maximum value of α_h is $n_1 n_2 n_3 \ldots$. If $p=0$ (no pseudotranslation occurs) (4) reduces to the classical Wilson statistics. (4) is in agreement with the observation (5,6) that the repetition of a subunit along a line introduces a modulation which increases the dispersion of the intensities. We say that F_h is a superstructure reflexion if $\alpha_h=0$: then $\langle |F_h|^2 \rangle = \epsilon \Sigma_q$. Otherwise F_h is a substructure reflexion. This definition does not always agree with definition given by previous authors. According to them (2,3,4,8) reflexions belonging to sets {A} or {B} are called substructure and superstructure reflexions respectively. A current notation denotes them "strong" and "weak" reflexions respectively, owing to the fact that the average value (4) for the set {A} is expected to be markedly larger than for the set {B}. Such definitions are not quite satisfactory in this context. Let us elucidate this statement by three connected examples.

(1) In $P222$ a pseudotranslation $u=⅓a + ¼c$ occurs, corresponding to the simultaneous presence of two independent pseudotranslations $u_1=a/3$ and $u_2=c/4$.

A subcell (see Fig. 1), the volume of which is 1/12 of the true cell, defines the set {A} whose points satisfy the condition $4h+3l=0$ (mod 12). According to (4), two values are allowed for a_h: $a_h=12$ for the set {A}, $a_h=0$ for the set {B}. In this example the subdivision of reciprocal space in substructure and superstructure reflexions given by Buerger finds a precise counterpart in our statistical results. Indeed two different values of a_h characterize the sets of substructure and superstructure reflexions respectively. The next two examples will show that it is not always true.

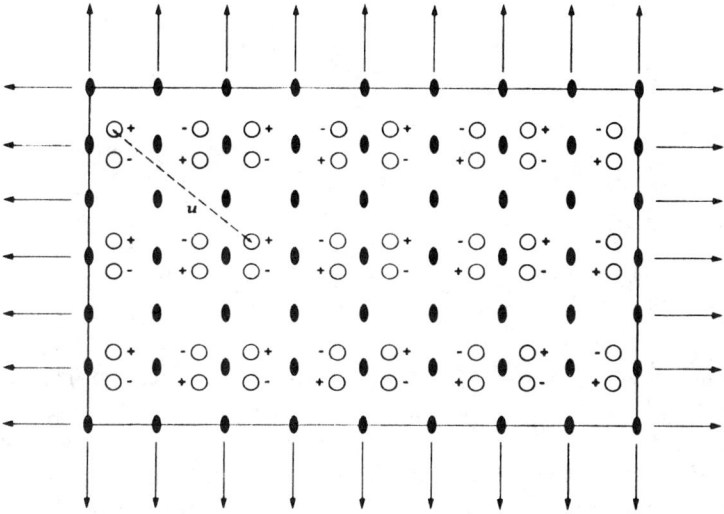

Figure 1. A pseudotranslation $u = \frac{1}{3}a + \frac{1}{4}c$ creates in a $P222$ cell 48 atomic sites, equivalent because of the symmetry or of u. The pseudocell is 1/12 of the volume of the original cell.

(2) Let the pseudotranslation $u_1=(a+b)/4$ occurs in $P2_12_12_1$. An additional symmetry dependent pseudotranslation $u_2=(-a+b)/4$ arises (see Fig. 2). A subcell $C222$ exists having the same periods as the $p+q$ atoms cell. The set {A} comprises points for which $h+k \equiv 0$ (mod 2) but it is not statistically homogeneous. Indeed: $a_h=4$ for reflexions satisfying

$$h+k \equiv 0 \text{ (mod 4) and } h-k \equiv 0 \text{ (mod 4)}, \qquad (7)$$

$a_h=2$ for reflexions for which

$$h+k \equiv 0 \text{ (mod 4) or } h-k \equiv 0 \text{ (mod 4)}. \qquad (8)$$

$a_h=0$ for substructure reflexions which do not satisfy (7) or (8), and for superstructure reflexions. We stress the fact that in contradiction with usual expectations, a subset of the set {A} corresponds to systematically weak reflexions.

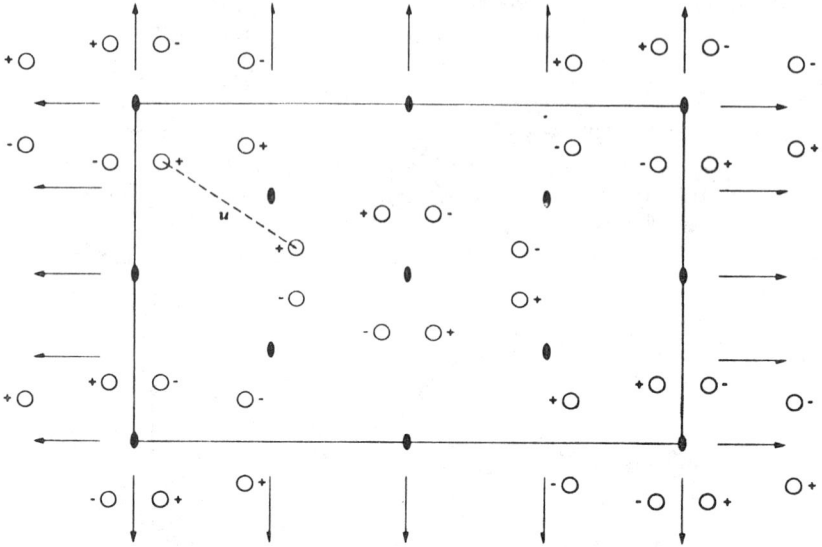

Figure 2. The pseudotranslation $u=(a+b)/4$ in $P222$ gives rise to a pseudocell $C222$ having additional non-crystallographic symmetry.

(3) Let the pseudotranslation $u_1=a/2$ occur in $P4$. Then an additional symmetry-dependent pseudotranslation $u_2=b/2$ arises. No subcell may be identified (see Fig. 3) so that the set {A} coincides with the full reciprocal lattice. In spite of that the pseudotranslation has sensitive effects on the intensity distributions. Indeed: $\alpha_h=2$ for reflexions with $h\equiv0$ (mod 2) and $k\equiv0$ (mod 2); $\alpha_h=1$ when $h\equiv0$ (mod 2) or $k\equiv0$ (mod 2); $\alpha_h=0$ otherwise.

The above examples show that Buerger's definitions sometimes obscure important statistical properties of reflexions. According to our approach a reflexion is classified as a superstructure reflexion if its intensity does not receive any contribution from the substructure or, more generally, from the p atoms suffering pseudotranslational symmetry. The above examples describe three typical situations: (i) non-crystallographic translational symmetry is completely contained in the crystallographic symmetry of the p-atom cell. This is the case of the example 1, where a non-conventional twelve-centered cell arises. Then $\epsilon_h\alpha_h$ is the statistical weight of the reflexion h for the p-atom cell; (ii) the translational symmetry is partially contained in the crystallographic symmetry of the p-atoms cell. In the example 2 the new space group $C222$ suffers a non-crystallographic translation responsible for the additional modulation of intensities; (iii) translational symmetry is unable to produce a substructure, and only modulates intensities. This is the case of example 3.

Recognition of Non-crystallographic Symmetry: Experimental

A-priori information, when available, makes easier the solution of several problems. *A-priori* information on pseudotranslational symmetry helps in solving crystal

Structure Factors & Pseudotranslations

structures because it allows a more correct definition of the primitive random variables in direct procedures.

Böhme (9) suggested that pseudotranslations can be recognized *via* the unequal distributions of the normalized structure factors when normalization is executed

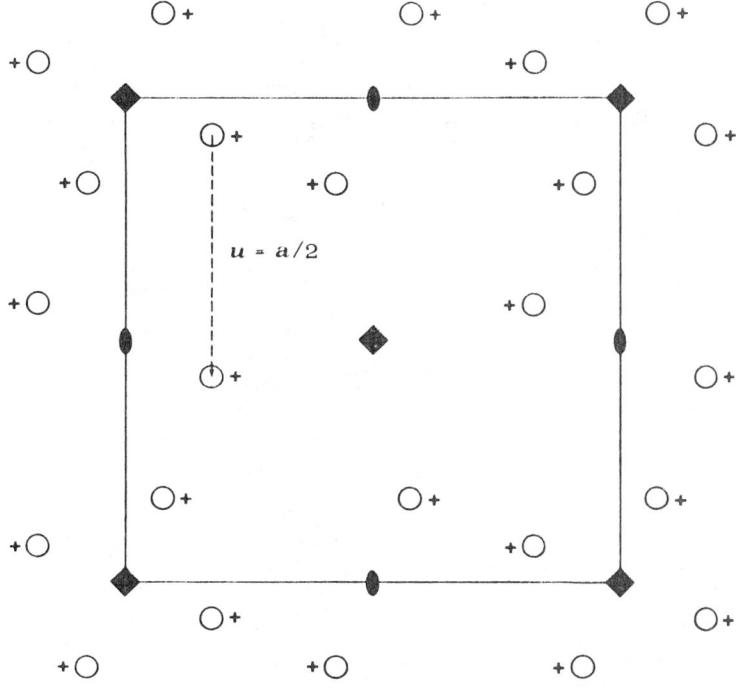

Figure 3. The pseudotranslation $u=a/2$ in a $P4$ cell does not create any pseudocell.

without taking translational symmetry into account. Let E' will denote such a pseudonormalized structure factor. In this view we selected the 64 different pseudotranslations having index $n<16$ provided $\mu'_i \leq 4$ for $i=1,2,3$. To each u the set of substructure reflexions (see eq. (6)) is associated: in practice the set for which

$$\frac{h v'_1}{\mu'_1} + \frac{k v'_2}{\mu'_2} + \frac{l v'_3}{\mu'_3} \equiv 0 \,(\text{mod } 1) \tag{9}$$

For each set the number of reflexions satisfying (9) and the average intensity are calculated.

If all the atoms have similar unitary scattering factor then

$$\langle |E'_h|^2 \rangle = (a_p \Sigma_p + \Sigma_q)/\Sigma_N \simeq (a_h p + q)/N \tag{10}$$

where

$$\Sigma_p = \sum_{j=1}^{p} f_j^2, \ \Sigma_N = \sum_{j=1}^{N} f_j^2.$$

From (10)

$$q/N \simeq (\langle|E'_h|^2\rangle - \alpha)/(1-\alpha), \qquad (11)$$

where the average is made by allowing h to vary over an homogeneous set of reflexions all having $\alpha_h = \alpha$. If the set coincides with the superstructure reflexions set, then $\alpha_h = 0$ and (11) reduces to

$$q/N \simeq \langle|E'_h|^2\rangle. \qquad (12)$$

(12) has been used in our calculations in order to estimate p and q for each of the 64 pseudotranslations, A figure of merit

$$\langle|E'|^2\rangle_{sub}/\langle|E'|^2\rangle_{sup} \qquad (13)$$

is calculated for each set: the actual pseudotranslation is expected to correspond to the set having the largest value of the ratio (13).

We have tested the procedure over a large variety of random and real structures in which only one pseudotranslation occurs, and always we obtained satisfactory indications. To give a numerical example, consider a randomly generated structure, space group $P\bar{1}$, $u = \frac{1}{2}a + \frac{1}{2}b + \frac{1}{3}c$, $t_p = 5$, $t_q = 15$, $p = 60$, $q = 30$. The $N(z)$ cumulative distribution, shown in Fig. 4, has an hypercentric character. In Table I the average values $\langle|E'|^2\rangle$ are given for the various 64 sets. Our procedure correctly selects the pseudotranslation $n.58$ and assigns $t_p=4.7$, $t_p=16.9$, in good agreement with reality. We applied the same procedure to intensities by freieslebenite (10), PbAgSbS$_3$, space group $P2_1/a$, $Z=4$. There are two pseudotranslations $u_1=a/2$, $u_2=b/3$ relating atoms of different type (Sb,Pb,Ag). In Table II the average $\langle|E'|^2\rangle$ are given for the 64 sets. Our procedure correctly selects pseudotranslation $n.53$ and assigns $p/N=0.55$, which is a quite reasonable value.

On the assumption that pseudotranslational symmetry has been correctly identified, the primitive random variables of our system can be brought up to date. That is to say, renormalization can be accomplished according to

$$E_h = F_h/\langle|F_h|^2\rangle^{1/2}$$

where $\langle|F_h|^2\rangle$ is given by (4). It is expected that distribution of the $|E|$'s will lose the hypercentric character and will approximate the centrosymmetric or non-centrosymmetric distribution according to whether the space group is centrosymmetric or not. In Figures 4 and 5 this behaviour is clearly shown.

Structure Factors & Pseudotranslations

Table I

A $P\bar{1}$ randomly generated structure: the average $\langle |E|^2 \rangle$ for 64 sets of reflexions. The program indicates the correct pseudotranslation by a double arrow.

#	expr	mod	⟨E**2⟩	#	expr	mod	⟨E**2⟩	#	expr	mod	⟨E**2⟩	#	expr	mod	⟨E**2⟩
1)	a l l	=	1.000	2)	h	= 2n	1.081	3)	k	= 2n	1.115	4)	l	= 2n	.845
5)	h+k+l	= 2n	.835	6)	h+k	= 2n	1.684	7)	h+l	= 2n	1.063	8)	k+l	= 2n	1.059
9)	h	= 3n	.989	10)	k	= 3n	.941	11)	l	= 3n	2.243	12)	h+k	= 3n	1.040
13)	h+l	= 3n	.974	14)	k+l	= 3n	.877	15)	h+k+l	= 3n	1.043	16)	h+k+2l	= 3n	1.029
17)	h+2k+l	= 3n	.970	18)	2h+k+l	= 3n	.961	19)	h+2k	= 3n	.999	20)	h+2l	= 3n	.988
21)	k+2l	= 3n	.881	22)	l	= 4n	1.281	23)	k	= 4n	.997	24)	h	= 4n	1.019
25)	h+k	= 4n	1.462	26)	h+l	= 4n	1.061	27)	k+l	= 4n	.944	28)	h+k+l	= 4n	.689
29)	2h+2k+l	= 4n	1.308	30)	2h+k+2l	= 4n	1.300	31)	h+2k+2l	= 4n	.970	32)	2h+k+l	= 4n	1.226
33)	h+2k+l	= 4n	1.010	34)	h+k+2l	= 4n	1.625	35)	h+2k	= 4n	1.008	36)	h+2l	= 4n	.999
37)	k+2l	= 4n	1.296	38)	2h+k	= 4n	.971	39)	2h+l	= 4n	.991	40)	2k+l	= 4n	.957
41)	3h+3k+l	= 4n	.604	42)	3h+k+3l	= 4n	.718	43)	h+3k+3l	= 4n	.687	44)	h+2k+3l	= 4n	1.049
45)	h+3k+2l	= 4n	1.439	46)	3h+k+2l	= 4n	1.439	47)	h+3k	= 4n	1.666	48)	h+3l	= 4n	.972
49)	k+3l	= 4n	1.222	50)	3k+2l	= 6n	2.488	51)	2k+3l	= 6n	.714	52)	2h+3k	= 6n	1.139
53)	3h+2k	= 6n	.926	54)	3h+2l	= 6n	2.456	55)	2h+3l	= 6n	.843	56)	2h+2k+3l	= 6n	.711
57)	3h+2k+3l	= 6n	1.069	58)	3h+3k+2l	= 6n	4.245 ⟪	59)	4k+3l	=12n	.980	60)	4h+3l	=12n	1.307
61)	4h+3k	=12n	1.286	62)	3k+4l	=12n	2.218	63)	3h+4k	=12n	1.248	64)	3h+4l	=12n	2.315

Table II

Freieslebenite: the average $\langle |E|^2 \rangle$ for 64 sets of reflexions. The program indicates the correct pseudotranslation by a double arrow.

#	expr	mod	⟨E**2⟩	#	expr	mod	⟨E**2⟩	#	expr	mod	⟨E**2⟩	#	expr	mod	⟨E**2⟩
1)	a l l	=	1.000	2)	h	= 2n	1.665	3)	k	= 2n	.960	4)	l	= 2n	1.057
5)	h+k+l	= 2n	.990	6)	h+k	= 2n	.917	7)	h+l	= 2n	1.031	8)	k+l	= 2n	.986
9)	h	= 3n	1.055	10)	k	= 3n	2.022	11)	l	= 3n	.976	12)	h+k	= 3n	.835
13)	h+l	= 3n	.970	14)	k+l	= 3n	.810	15)	h+k+l	= 3n	.787	16)	h+k+2l	= 3n	.803
17)	h+2k+l	= 3n	.787	18)	2h+k+l	= 3n	.803	19)	h+2k	= 3n	.835	20)	h+2l	= 3n	.964
21)	k+2l	= 3n	.810	22)	l	= 4n	1.051	23)	k	= 4n	.923	24)	h	= 4n	1.685
25)	h+k	= 4n	.707	26)	h+l	= 4n	1.024	27)	k+l	= 4n	.964	28)	h+k+l	= 4n	.989
29)	2h+2k+l	= 4n	1.070	30)	2h+k+2l	= 4n	.966	31)	h+2k+2l	= 4n	2.402	32)	2h+k+l	= 4n	.965
33)	h+2k+l	= 4n	1.012	34)	h+k+2l	= 4n	.680	35)	h+2k	= 4n	1.664	36)	h+2l	= 4n	1.618
37)	k+2l	= 4n	.958	38)	2h+k	= 4n	.852	39)	2h+l	= 4n	1.051	40)	2k+l	= 4n	1.084
41)	3h+3k+l	= 4n	.979	42)	3h+k+3l	= 4n	.989	43)	h+3k+3l	= 4n	.979	44)	h+2k+3l	= 4n	1.000
45)	h+3k+2l	= 4n	.680	46)	3h+k+2l	= 4n	.680	47)	h+3k	= 4n	.707	48)	h+3l	= 4n	1.027
49)	k+3l	= 4n	.964	50)	3k+2l	= 6n	.926	51)	2k+3l	= 6n	2.107	52)	2h+3k	= 6n	.949
53)	3h+2k	= 6n	3.684 ⟪	54)	3h+2l	= 6n	1.631	55)	2h+3l	= 6n	1.104	56)	2h+2k+3l	= 6n	.883
57)	3h+2k+3l	= 6n	2.081	58)	3h+3k+2l	= 6n	.896	59)	4k+3l	=12n	2.057	60)	4h+3l	=12n	1.079
61)	4h+3k	=12n	.842	62)	3k+4l	=12n	.852	63)	3h+4k	=12n	3.685	64)	3h+4l	=12n	1.593

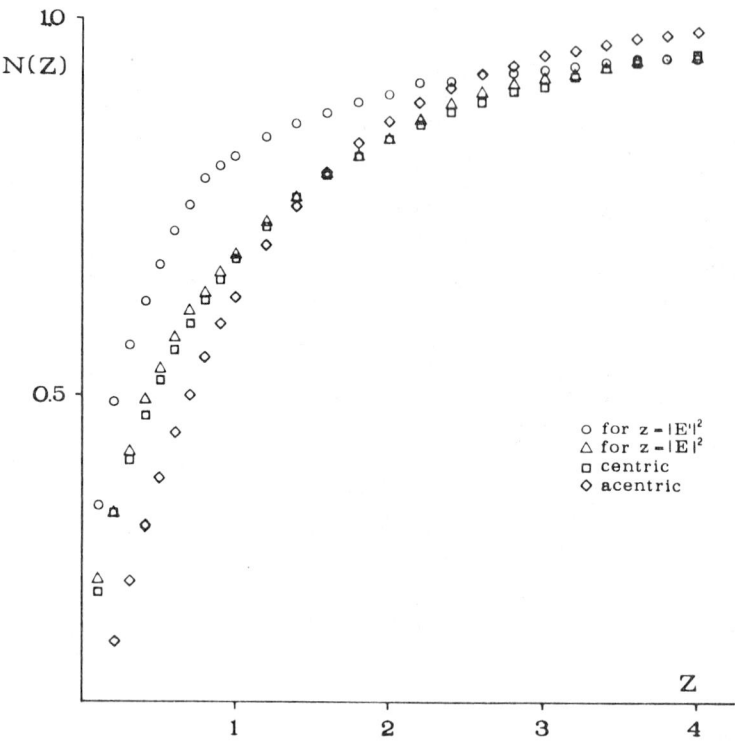

Figure 4. Cumulative distributions $N(z)$ for a $P\bar{1}$ randomly generated structure before and after renormalization, compared with theoretical centro- and non-centrosymmetrical distributions.

Conclusions

The mathematical model so far described provides a reasonable tool for handling, from a statistical point of view, the effects of non-crystallographic symmetry on diffraction intensities. Even if the model seems valid and provides useful information in a sufficiently large number of situations, it suffers the following limits:

(1) if t_p (and/or t_q) is too small then the central limit theorem, which is at basis of the present mathematical treatment, does not strictly hold. In particular, the p and q values arising from the statistical analysis of the intensities may markedly be affected by this fact. We have made specific tests over a large variety of crystallographic situations characterized by $t_p=1,2,...$. The results suggest that the approach usually works well also in this critical situations.

(2) When more pseudotranslational vectors u_i are contemporaneously present we assumed that the same t_p atoms contemporaneously suffer the various u_i. That is usually but not always true.

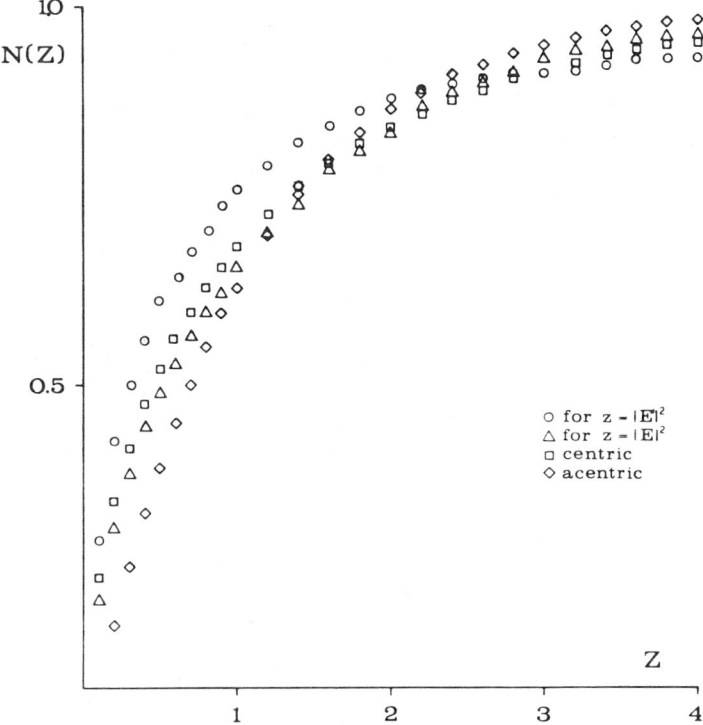

Figure 5. Cumulative distribution $N(z)$ for freieslebenite before and after normalization, compared with theoretical centro- and non-centrosymmetrical distributions.

(3) So far all the p atoms are assumed to be present in the structure. Not always that occurs: indeed some of the p atoms may be missed.

In conclusion, more efforts are necessary for handling more complex crystallographic situations. However a quite satisfactory solution of the problem may not exist. Even in spite of the above limits the present approach is largely applicable and proved useful in direct methods of solving crystal structures (11).

References and Footnotes

1. *Present address:* Rudjer Bosković Institute, Bijenicka 54, 41000 Zagreb, Yugoslavia.
2. M. Buerger, *Proc. Natl. Acad. Sci. USA, 42,* 776-781, 1956
3. M. Buerger, *Vector space,* New York, John Wiley, 1959
4. K. Taxer, *Z. Kristallogr., 155,* 1-25, 1981
5. H. Lipson and M. M. Woolfson, *Acta Cryst. 5,* 680-682, 1952
6. D. Rogers and A.J.C. Wilson, *Acta Cryst. 6,* 439-449, 1953
7. G. Cascarano, C. Giacovazzo and M. Luić. In preparation
8. J.W. Jeffery, *Acta Cryst. 17,* 776-777, 1964
9. R. Böhme, *Acta Cryst. A38,* 318-326, 1982
10. T. Ito and W. Novacki, *Z. Kristallogr. 139,* 85-102, 1974
11. G. Cascarano, C. Giacovazzo and M. Luić. In preparation

Structure & Statistics in Crystallography, ISBN 0-940030-10-1,
Ed., A. J. C. Wilson F. R. S., Adenine Press, ©Adenine Press 1985.

A Best Test to Distinguish Between X-ray Intensity Distributions for Space-group Determination

S. Parthasarathy and N. Elango
Department of Crystallography & Biophysics
University of Madras
Madras—600 025, India

Abstract

A statistical analysis of X-ray intensities of a crystal can provide information regarding crystal symmetry (*e.g.* the presence or absence of a centre of symmetry) and hence can be utilised for space group determination. In this connection a crystallographer needs an efficient method of testing whether the normalized intensity data of a given crystal are a random sample from a distribution defined by the probability density function $P_0(z)$ rather than from an alternative distribution defined by the probability density function $P_1(z)$. A best test (also called a most powerful test) can be constructed for this purpose by invoking the Neyman-Pearson lemma. In this paper the general procedure for doing this is described. The method is then applied to construct a best test to distinguish between the centric and acentric Wilson distributions. The method has been tested in the case of a few crystals.

1. Introduction

A problem of interest in connection with statistical tests for space group determination consists in finding whether the normalized intensity data of a given crystal are a random sample from one or the other of two explicitly known distributions. For example, one may be interested in testing whether the given normalized intensity data is a random sample from either the centric Wilson distribution or the acentric Wilson distribution. Though such problems can be tackled by using different statistical criteria (1), yet statistically the best way of doing this consists in stating the problem on hand as a problem of testing a simple null hypothesis versus a simple alternative hypothesis and tackling it by making use of the Neyman-Pearson lemma (2, 3). In this paper we shall describe the general procedure involved in constructing such a best test. We shall illustrate the method by constructing a best test for distinguishing between the acentric and centric Wilson distributions. We shall also take into account the fact that the intensity data of an actual crystal are subject to a truncation at low values of the intensity due to unobserved reflexions. The results of applying the present method to a few actual crystals are also given.

2. Notation and Results from Probability Theory

Suppose that $x_1, x_2, ..., x_n$ is a random sample of size n from a population whose probability distribution is characterized by the probability density function (*pdf*, hereafter) $P(x)$. Suppose that we know $P(x)$ could be either the *pdf* $P_0(x)$ or the *pdf* $P_1(x)$. We shall assume that $P_0(x)$ and $P_1(x)$ are known explicitly so that their values can be computed for any given value of x. The hypothesis testing problem may be stated as

$$H_0 : P(x) = P_0(x)$$
$$H_1 : P(x) = P_1(x)$$
(1)

where H_0 and H_1 are the null and alternative hypotheses respectively. Let α and β respectively be the type-I and type-II errors of the test and let T be the statistic used in the test. The statistic partitions the n-dimensional sample space into two disjoint regions called the region of acceptance and the *critical region*. The critical region of the test is denoted by C and it is the part of the sample space that corresponds to the rejection of the null hypothesis H_0. The probability α that the sample point $(x_1, x_2, ..., x_n)$ lies in the critical region when H_1 is true is called the power of the test. For the hypotheses H_0 and H_1 and given values of α and n the critical region that maximizes the power $1-\beta$ (or equivalently that which minimizes the type-II error β) is called a *best critical region* (*BCR*, say). A test based on a *BCR* is defined to be a *best test* or a *most powerful test*.

A best test can be constructed by making use of the Neyman-Pearson lemma which may be stated as follows:

Let $L_i(x_1, x_2, ..., x_n)$, $i = 0, 1$, be the likelihood function of the sample $x_1, x_2, ..., x_n$ corresponding to the hypotheses H_0 and H_1 respectively. If there exists a critical region C of size α and a non-negative constant k such that

$$\frac{L_1(x_1, x_2, ..., x_n)}{L_0(x_1, x_2, ..., x_n)} \geq k \text{ for points in } C$$

and

$$\frac{L_1(x_1, x_2, ..., x_n)}{L_0(x_1, x_2, ..., x_n)} \leq k \text{ for points not in } C$$

then C is a *BCR* of size α.

3. Method of Constructing a Best Test to Distinguish between Two Intensity Distributions

Suppose that we want to know whether the normalized intensity data $z_1, z_2, ..., z_n$ of a given crystal is a random sample from either a distribution with the *pdf* $P_0(z)$ or

Intensity & Space-group Determinations

that with the *pdf* $P_1(z)$. We shall take $P_0(z)$ and $P_1(z)$ to define the hypotheses H_0 and H_1 respectively. That is,

$$H_0 : P(z) = P_0(z)$$
$$H_1 : P(z) = P_1(z) \qquad (2)$$

where $P(z)$ is the *pdf* of the distribution of z for the given crystal. From Eq. (2) we obtain the likelihood functions for the sample corresponding to the hypotheses H_0 and H_1 to be

$$L_i(z_1, z_2, \ldots, z_n) = \prod_{j=1}^{n} P_i(z_j), \; i = 0, 1 \qquad (3)$$

From Eq. (3) we obtain the likelihood ratio for the sample to be

$$\frac{L_1(z_1, z_2, \ldots, z_n)}{L_0(z_1, z_2, \ldots, z_n)} = \prod_{j=1}^{n} \left[\frac{P_1(z_j)}{P_0(z_j)}\right] \qquad (4)$$

In view of the Neyman-Pearson lemma (see Section 2) and Eq. (4) the *BCR* will be determined by the inequality

$$\prod_{j=1}^{n} \left[\frac{P_1(z_j)}{P_0(z_j)}\right] \geq k \qquad (5)$$

where k is a non-negative constant. For the situation where $P_0(z)$ and $P_1(z)$ are sufficiently simple the l.h.s of Eq. (5) may be simplified so that Eq. (5) may be rewritten in an equivalent form as

$$T(z_1, z_2, \ldots, z_n) \geq a \qquad (6)$$

where T is a simple function of the sample $z_1, z_2, \ldots, z_n$ and a is a real number. Let $P(T \mid H_0)$ denote the *pdf* of the statistic T when H_0 is true. The *BCR* of size α will be given by the inequality

$$T \geq T_0 \qquad (7)$$

where the value of T_0 is such that

$$P(T \geq T_0 \mid H_0) = \alpha \qquad (8)$$

Since the l.h.s of Eq. (8) represents the complementary cumulative function of T, we can rewrite Eq. (8) compactly as

$$N_1^c(T_0) = \alpha \qquad (9)$$

Thus a best test for the present situation is the one that is based on the random variable T with the interval $T \geq T_0$ as the critical region. For a given α, the value of T_0 satisfying Eq. (8) can be obtained by the Monte Carlo method. If T' is the calculated value of the statistic T for the normalized intensity data $z_1, z_2, \ldots, z_n$ of the given crystal, then the data can be taken to be (at the significance level α) a random sample from the distribution $P_0(z)$ if $T' < T_0$ and from the distribution $P_1(z)$ if $T' > T_0$.

In Section 4 we shall illustrate the theoretical procedure by constructing a best test for distinguishing between the centric and acentric Wilson distributions.

4. A Best Test to Distinguish between the Centric and Acentric Wilson Distributions

Since data truncation arising from unobserved reflexions is an inherent property of the observed intensity data it is essential to take this into account in constructing a best test. If z_t is the threshold of observation for the normalized intensity (4), then the acentric Wilson distribution applicable to the truncated data (*i.e.* the data for which $z_t \leq z < \infty$) can be written as (5)

$$P_a(z) = c_0 \exp(-z) \tag{10}$$

where c_0 is defined to be

$$c_0 = \exp(z_t) \tag{11}$$

The centric Wilson distribution applicable to the truncated data can be written as (5)

$$P_c(z) = c_1 z^{-\frac{1}{2}} \exp(-z/2) \tag{12}$$

where c_1 is defined to be

$$c_1 = (2\pi)^{-\frac{1}{2}} / [1 - \mathrm{erf}(\sqrt{z_t/2})] \tag{13}$$

We can state the present hypotheses testing problem as

$$\begin{aligned} H_0 &: P(z) = c_0 \exp(-z) \\ H_1 &: P(z) = c_1 z^{-\frac{1}{2}} \exp(-z/2) \end{aligned} \tag{14}$$

From Eq. (5) and Eq. (14) it is seen that the *BCR* will be determined by the inequality

$$\left(\frac{c_1}{c_0}\right)^n \exp(\tfrac{1}{2} \sum_{i=1}^n z_i) \prod_{i=1}^n (z_i^{-\frac{1}{2}}) \geq k \tag{15}$$

where k is some non-negative constant. Taking natural logarithms the equivalent inequality obtained from Eq. (15) is

Intensity & Space-group Determinations

$$\sum_{i=1}^{n} z_i - \sum_{i=1}^{n} \log z_i \geq 2[\log k - n\log(c_1/c_0)] \tag{16}$$

We can write

$$\sum_{i=1}^{n} z_i - \sum_{i=1}^{n} \log z_i = n\langle z - \log z \rangle$$
$$= nT, \text{ say} \tag{17}$$

where T is defined to be

$$T = \langle z - \log z \rangle \tag{18}$$

Making use of Eq. (17) we can rewrite Eq. (15) as

$$T \geq a$$

where a is a real number defined by

$$a = (2/n)[\log k - n\log(c_1/c_0)] \tag{19}$$

For a given α, if T_0 is the solution of Eq. (8), then the BCR of size α will be given by the inequality $T \geq T_0$. We shall presently consider the method of evaluating T_0 for a given α.

Since it is too difficult to calculate the complementary cumulative function $N_0^c(T)$ by an analytical method we shall obtain it by the Monte Carlo method. For this we shall make use of the principle of the probability integral transformation which may be stated as follows (6): If $N(x)$ is the cumulative function of a random variable x, then the random variable u defined by

$$u = N(x) \tag{20}$$

has a uniform distribution in the interval $0 \leq u \leq 1$.

Since the cumulative function of z under H_0 is known to be (5)

$$N(z) = 1 - c_0\exp(-z), \, z_t \leq z < \infty \tag{21}$$

the random variable u defined by

$$u = 1 - c_0\exp(-z) \tag{22}$$

would have a uniform distribution in the interval $0 \leq u \leq 1$. We can rewrite Eq. (22) as

$$z = \log[c_0/(1-u)] \tag{23}$$

We may use a random number generator to obtain a sequence of m random numbers which may be treated as a random sample of size m from a uniform distribution and use these in Eq. (23) to obtain m values of z. Using these z-values we may obtain the value of T from Eq. (18). This procedure may be repeated with different random number sequences to generate a large number (r, say) of values of T. The T-values thus obtained may be treated as a random sample of size r from the required distribution $P(T \mid H_0)$. The complementary cumulative function $N_0^c(T)$ may then be obtained from these values of T. The solution of Eq. (8) for a given α (0.05, say) may then be obtained by linear interpolation. Since $P_0(z)$ and $P_1(z)$ depend on the truncation limit $z_t(=y_t^2)$, the value of T_0 for a given value of α will also be a function of y_t. The solutions of Eq. (8) for $\alpha = 0.05$ were obtained corresponding to different values of y_t and these are given in Table 1.

Table I
Value of T_0 defining *BCR* of size 0.05 in a best test for distinguishing between the centric and acentric Wilson distributions as a function of the truncation limit y_t

y_t	0.05	0.10	0.15	0.20	0.25	0.30	0.35	0.40	0.50
T_0	1.620	1.589	1.552	1.514	1.476	1.439	1.405	1.375	1.330

5. Discussion of the Theoretical Results

The following points regarding the present method may be noted: (i) In Section 4 we have illustrated the method of constructing a best test for a particular situation, namely, that of distinguishing between the acentric and centric Wilson distributions. A similar procedure may be applied to other situations provided the functions $P_0(z)$ and $P_1(z)$ corresponding to the hypotheses H_0 and H_1 are such that the statistic T takes a simple form. Otherwise the procedure may turn out to be theoretically as well as numerically too complex (7). (ii) The advantage of the present method is that it is well suited for taking automatic decision by digital computer since the final decision requires only a single value of the statistic T to be calculated from the individual values of the normalized intensity data of the crystal.

6. Test of the Method

The theoretical result obtained in Section 4 was tested in the case of a few structures and the details are given in Table 2. In the case of each crystal the value of the statistic T was computed from the calculated normalized intensity data. The reflexions for which the normalized structure factor magnitude $y > 0.2$ (*i.e.* $z > 0.04$) were used in the calculations. In the case of crystals (1) and (2) of Table 2 reflexions for which $0.2 \leq \sin\theta/\lambda \leq 0.5\text{Å}^{-1}$ were used in the test while in the case of crystal (3) the reflexions for which $0.2 \leq \sin\theta/\lambda \leq 0.6\text{Å}^{-1}$ were used. It is seen that the test leads to correct results in all the cases.

Intensity & Space-group Determinations

Table II
Test of the method in a few crystals

No.	Crystal	Asymemtric Unit	Reference	Space group	N	T'	T_0	Conclusion
1.	Dimethyl ester of meso tartaric acid	C_6O_6	(8)	$P\bar{1}$	600	1.671	1.514	C
2.	2.2.2. (1,3,5) cyclophane 1,9,17 triene	$C_{18}H_{12}$	(9)	$P\bar{1}$	931	1.869	1.514	C
3.	L-N-acetyl histidine monohydrate	$2(C_8O_3N_3H_{11})$	(10)	$P1$	1705	1.480	1.514	NC

Note: N is the number of reflexions used in the test. $T \geq T_0$ is the critical region of the test.

Acknowledgements

One of the authors (N. E.) wishes to thank University Grants Commission (India) for financial assistance.

References and Footnotes

1. Ponnuswamy, M. N. and Parthasarathy, S. *Acta Cryst. A33*, 838-844 (1977).
2. Beaumont, G. P. *Elementary Mathematical Statistics,* McGraw Hill Book Company (UK) Limited (1972).
3. Hoel, P. G., Port, S. C. and Stone, C. J. *Introduction to Statistical Theory,* Houghton Mifflin Company, Boston (1971).
4. We shall also write $z_t = y_t^2$ where y_t is the threshold value of the normalized structure factor magnitude.
5. Ponnuswamy, M. N. *Ph.D. Thesis.* Univ. of Madras, India (1979).
6. Meyer, P. L. *Introductory Probability and Statistical Applications,* Addison-Wesley Publishing Company (1973).
7. Best tests to distinguish between other pairs of intensity distributions are under investigation.
8. Kroon, J. and Kanters, J. A. *Acta Cryst. B29,* 1278-1283 (1973).
9. Hanson, A. W. and Rohrl, M. *Acta Cryst. B28,* 2287-2291 (1972).
10. Kistenmacher, J. J., Hunt, D. J. and Marsh, R. H. *Acta Cryst. B28,* 3352-3361 (1972).

Structure & Statistics in Crystallography, ISBN 0-940030-10-1,
Ed., A. J. C. Wilson F. R. S., Adenine Presss, ©Adenine Press 1985.

Effect of Statistical Fluctuations and Systematic Errors on Intensity Distributions—Which Way Forward?

A.J.C. Wilson
Crystallographic Data Centre
University Chemical Laboratory
Cambridge CB2 1EW, England

Abstract

Depending on the assumptions made, one or more of the following representations of an ideal probability distribution of intensities of reflexion (or structure factors) may be practicable:

(1) An expression in closed form.
(2) A series expansion in orthogonal polynomials.
(3) A Fourier (Barakat) series.
(4) A steepest-descents approximation.

[For recent references see A.J.C. Wilson, *Proc. Ind. Acad. Sci. Chem. Sci.*, *92.*, 335-339, 1983.] Observed distributions, even if the underlying distribution is ideal, are distorted by statistical fluctuations and systematic errors in the intensity measurements, and a valid comparison of a calculated and an observed distribution must take these distortions into account. In principle the distribution of the statistical fluctuations will be fully predictable [for a discussion of several cases see A.J.C. Wilson, *Acta Crystallogr.*, A *36*, 929-936, 1980], but little is known of the distribution of systematic errors—often no more than the first and second moments, sometimes even less. An attempt is made to modify the representations in the light of limited knowledge of the fluctuations and errors. Closed forms have been obtained only in simple cases; known moments can be incorporated in orthogonal-polynomial expressions, but convergence is slow if atomic heterogeneity is high; Fourier representations seem unpromising; steepest-descents representations have not been developed as yet.

Introduction

The purpose of this paper is to discuss the effects of statistical fluctuations and of systematic errors on the observed probability distributions of intensities. Systematic errors may blur the details of the distribution function, and may shift its position; random errors blur its details but do not alter its mean. An example of the systematic type is uncorrected extinction, which shifts the higher intensities systematically to lower values. An example of a random error is statistical fluctuations in counting rates, which result in shifts that are randomly to higher or to lower values.

The distribution is thus broadened, the higher parts being lowered and the lower raised, but its centre of gravity is not affected (1).

When only the intrinsic probability distributions are being considered, it does not greatly matter whether the variable chosen is the intensity of reflexion (I), or its positive square root, the modulus of structure factor (F), since both are necessarily real and non-negative. In an obvious notation, the relation between the expressions for the intensity distribution and the structure-factor distribution is

$$P_I(I) = \tfrac{1}{2} I^{-\frac{1}{2}} p_F(I^{\frac{1}{2}}) \tag{1}$$

or

$$p_F(F) = 2F p_I(F^2) . \tag{2}$$

Statistical fluctuations in the counting rates, however, introduce a small but finite probability of negative observed intensities [(3), (4)], and thus of imaginary 'observed' structure factors.

Four representations of probability distributions of intensities or structure factors have been proposed:

1. Closed-form expressions;
2. Expansions in orthogonal polynomials;
3. Fourier series;
4. Asymptotic or steepest-descents approximations.

The literature [for recent references see (5)] contains many discussions of both p_F and p_I for the first and second of these, but the third and fourth seem to have been applied to p_F only.

As already mentioned, the distribution of intensities actually observed is not simply that inherent in the atomic arrangement, but is modified ('smeared') by the statistical fluctuations in the counting rates [see, for example, (6); (7), section 1.6; (4)]. Probability distributions can be derived for the fluctuations; in the simplest case the distribution is Poisson: if the ideal number of counts that should be observed in the fixed counting-time interval is N, the probability of actually observing N_o is

$$p(N_o) = (N_o!)^{-1} N^{N_o} \exp(-N) . \tag{3}$$

The Poisson distribution holds only if the background is negligible. If the background is appreciable there is a finite possibility of N_o being negative [(4), (5)], with a probability distribution

$$p(N_o) = \exp\{-(B+T)\}(T/B)^{\frac{1}{2}N_o} I_{N_o}\{2(BT)^{\frac{1}{2}}\} , \tag{4}$$

where T and B are the numbers of counts that should be observed when the counter is set to receive (reflexion including background) and (background only), and I_n is the modified Bessel function of order n. If the counting times are different for reflexion and background, or if fixed-count timing is used instead of fixed-time counting, the distribution of N_o becomes even more complex; examples and some low-order moments have been given by Wilson (4). It is frequently assumed that the true distribution can be approximated by a Gaussian:

$$p(I_o) = (2\pi\sigma^2)^{-\frac{1}{2}}\exp\{-(I_o-I)^2/2\sigma^2\}, \qquad (5)$$

where I is the 'true' intensity and σ is the standard deviation of I_o. Ideally, σ should be simply related to the number of counts used in the determination of I, but in practice semi-empirical modifications are made.

Closed-Form Combined Distributions for Simple Cases

The simplest case of a combined distribution is obtained when the structural distribution has the ideal acentric form (6):

$$p(I)dI = \Sigma^{-1}\exp(-I/\Sigma)dI \qquad (6)$$

and the fluctuation distribution has the Poisson form [equation (3)]. Physically, the Poisson assumption corresponds to negligible background. In this case the probability of observing an intensity of N_o counts is

$$p(N_o) = (N_o!\Sigma)^{-1}\int_0^\infty \exp(-N)N^{N_o}\exp(-N/\Sigma)dN \qquad (7)$$

$$= \Sigma^{N_o}(\Sigma+1)^{-N_o-1}, \qquad (8)$$

where I is expressed in counts per time interval (N), and Σ also is expressed in counts per time interval. Though N_o is necessarily an integer, there is no such requirement for N and Σ.

When the structural distribution has the ideal centric form:

$$p(I)dI = (2\pi\Sigma)^{-\frac{1}{2}}I^{-\frac{1}{2}}\exp(-I/2\Sigma)dI \qquad (9)$$

and the statistical distribution has the Poisson form (1), the combined distribution takes the form

$$p(N_o) = (2\pi\Sigma)^{-\frac{1}{2}}(N_o!)^{-1}\int_0^\infty N^{(N_o-\frac{1}{2})}\exp[-N(1+2\Sigma)/2\Sigma]dN, \qquad (10)$$

where again I and Σ are expressed in counts per time interval. Integration gives

$$p(N_o) = (2\pi\Sigma)^{-\frac{1}{2}}(N_o!)^{-1}[2\Sigma/(1+2\Sigma)]^{N_o+\frac{1}{2}}\Gamma(N_o+\frac{1}{2}),\qquad(11)$$

where Γ is the gamma function.

The distributions (6) and (9) have the mean Σ and the variances Σ^2 and $2\Sigma^2$ respectively. Equation (8) represents a geometric distribution [(19), formula 28.1.24] with parameter $p = 1/(\Sigma+1)$, mean Σ and variance $\Sigma^2 + \Sigma$. The distribution (11) seems not to have a recognized name. However, the device of repeated differentiation with respect to a parameter indicates that it has the mean Σ and the variance $2\Sigma^2 + \Sigma$. As was to be expected, the statistical fluctuations described by the Poisson distribution have not affected the mean intensity of reflexion, but have increased the variance of the ideal acentric and centric distributions by an amount Σ.

The distribution of observed intensities when background is appreciable, as given in equation (4), is curious. It contains two parameters, B and T, in three of the four arithmetic combinations; the sum, ratio and product are there, but the difference, the 'true' intensity that it would be most convenient to have, is missing. One can, of course, express the sum, ratio, and product in terms of the difference $T - B$ and T or B or $T + B$ as an additional parameter, but the expressions do not combine readily with either of the structural distributions (6) and (9).

The normal distribution of errors [equation (5)] and the acentric distribution [equation (6)] can be combined without difficulty to give a closed-form expression in terms of the complementary error function, but the result is not attractive.

Expansions in Orthogonal Polynomials

Probability distributions of structure factors were originally derived on the basis of the central-limit theorem of statistics; under fairly general conditions, the sum of a number of independent random variables has an approximately normal distribution, with a mean equal to the sum of the means of the variables and a variance equal to the sum of the variances of the variables. The theorem applies separately to the real and the imaginary parts of the structure factor, and was applied by Wilson to determine the ideal limiting distributions for centrosymmetric crystals ['centric' distribution, (6)], non-centrosymmetric crystals ['acentric' distribution, (6)] and crystals that exhibit appreciable dispersion ['subcentric' distribution, (7)]. 'Hypercentric' (8, 9, 10) and 'sesquicentric' (11) distributions have also been considered, and the subcentric distribution applies also when a non-centrosymmetric crystal contains centrosymmetric groups (12).

The ideal distributions obtained by the central-limit theorem, by the very statement of the theorem, have the same mean and variance as those of the real distributions, but the theorem does not guarantee that the higher moments will agree. All that is promised is that disagreements will become less and less important as the number

Fluctuations & Errors in Intensity Distribution

of atoms increases, the atomic proportions being maintained. The ideal distributions may thus be poor representations if the number of atoms in the unit cell is small, especially if they differ widely in scattering power. An obvious way of improving the agreement is to modulate the ideal distribution by a correction factor that will force agreement of as many as desired of the higher moments. For structure factors the probability of $-F$ is the same as the probability of $+F$, so it is not necessary to consider odd moments in this case. The correction factor may thus be written

>1 + constant ×
>
>(difference of ideal and real fourth moments) ×
>
>(orthogonal polynomial of order four)
>
>+ appropriate terms in sixth, eighth, . . . moments if necessary.

The appropriate polynomials are Hermite for approximately centric distributions and Laguerre for approximately acentric (13,14), and unnamed but known for approximately subcentric (15). When the expressions are transformed into the intensity case no new moments of the F's appear.

The effect of experimental errors on a distribution function $p(x)$ may be described as follows. Each normalized structure amplitude that ought to have the value x has a certain probability, say $g(e)$, of being measured as $x + e$. In the usual convolution argument [see, for example, (16), pp. 188-192, or (in a different crystallographic context), (17), pp. 5-6, 82-84], the modified distribution of x is given by

$$p_1(x) = \int p(x-e)g(e)de , \qquad (12)$$

the integration being over all values of e for which $g(e)$ is appreciable.

The moments of such a convolution are easily estimated. In equation (11) replace x by z, where

$$z = x - e , \qquad (13)$$

and z is treated as a constant during the integration over e. The kth moment of the convolution is then

$$M_k = \iint x^k p(x-e)g(e)dedx \qquad (14)$$

$$= \iint (z+e)^k p(z)g(e)dedz \qquad (15)$$

$$= \Sigma [l!(k-l)!/k!] \int z^{k-l} p(z)dz \int e^l g(e)de \qquad (16)$$

$$= \Sigma [l!(k-l)!/k!] P_{k-l} G_l , \qquad (17)$$

where P_{k-l} is the $(k-l)$th moment of $p(x)$ and G_l is the lth moment of $g(e)$. It would therefore seem easy to incorporate all moments of $p(x)$ and $g(e)$ into the representation of the probability distribution of the intensities or structure factors. Unfortunately, in the present application, $g(e)$ will contain x as a parameter—equations (1) and (2) contain N explicitly, and equation (3) contains σ, which must be regarded as a function of I. The integrations in equation (15), therefore, no longer carry through. It may be argued, however, that for x (intensity or structure factor modulus) not too small, the range of e for which $g(e)$ is appreciable is small compared with x, so that z can be regarded as a constant during the integration over e. Equation (15) thus becomes

$$M_k = \sum [l!(k-l)!/k!] \int z^{k-l} G_l(z) p(z) \mathrm{d}z ,\qquad(18)$$

so that M_k can be evaluated if $g(e)$ is known as a function of z (which does not differ significantly from x under the assumption that x is large compared with the effective range of e). The moments of $p(x)$, the ideal distribution, are known up to about $l = 10$ [(14), (15)], and at least the first two moments of $g(e)$ are estimated in most structure determinations. The polynomial approximations to the resultant distributions can them be written down, making use of what moments are known (18).

In much structure determination the scaling of observed intensities or structure factors is done by making the average observed value equal to the average calculated value in each of successive ranges of $(2\sin\theta)/\lambda$; this has the effect of making the actual average error $\langle e \rangle$ very close to zero. At least two estimates of the variance of e are possible. First, it may be assumed that it arises only from statistical fluctuations in the numbers of counts recorded in the diffractometer measurements. Thus in fixed-time counting the statistical variance should be (in counts per time interval) the total number of counts used in determining the intensity of the reflexion, $B + T$, but in practice adjustments are made, which allow, at least in part, for other reasons for variations in the counts recorded. Secondly, it may be assumed that the differences between observed and calculated intensities (or structure factors) arise only from the random experimental errors, so that

$$\sigma^2 = \langle (x_o - x_c)^2 \rangle ,\qquad(19)$$

the mean being taken over a small range of values of x_i including the value of x (I or F) for which σ^2 is desired. The true value of variance may well lie between these two estimates; the first is certain to be low, and the second will include some systematic error as well as random fluctuation. In practice the variance of an intensity I is often approximated as

$$\sigma_I^2 = a + bI + cI^2 ,\qquad(20)$$

or of a structure factor by a similar equation in F. In principle a and b would be determined by B and T for each reflexion. In practice they, as well as c, may be regarded as semi-empirical parameters and adjusted to improve the agreement index R. It is rare for any attempt to be made to estimate moments higher than the variance. When no higher moments are available the approximation

$$p_1(x) = \tfrac{1}{2}[p(x - m + \sigma) + p(x - m + \sigma)], \qquad (21)$$

where m and σ are the mean and standard deviation of $g(e)$, may be useful. It incorporates all the available information about $g(e)$ and gives the correct mean and variance for x.

Fourier and Steepest-Descents Representations

The modulus of the structure factor of a reflexion has a precisely limited range: it cannot be less than zero or greater than Φ, the sum of the moduli of the structure factors. The intensity is similarly limited to the range from 0 to Φ^2. A Fourier series with a range equal to that of the possible values of F or I, therefore, has considerable attractions for representing the probability distributions of F or I. In practice, the Fourier series (for space group $P\bar{1}$) and the Fourier-Bessel series (for space group $P1$) developed by Weiss and co-workers [(20), (21), (22)] converge rapidly and give 'exact' representations of the ideal probability distributions of F. Fourier representations of the ideal distributions of I have not been investigated as yet.

Fourier methods do not seem promising for distributions modified by statistical fluctuations. The possible range of I is no longer finite, so the series loses its aesthetic attraction. There is a small but not negligible probability that intensities will be observed as negative [(3), (4)]; for a distribution of I these difficulties can be coped with by using a Fourier integral instead of a series. A negative observed intensity corresponds to a pure imaginary F, so two representations would be needed, one for F real, and one for F imaginary—not a pleasing prospect.

The effect of fluctuations and errors on steepest-descents representations has not been investigated.

Summary

It must be concluded, regretfully, that there is no obvious easy way of representing distributions modified by random or systematic errors. Orthogonal-polynomial expansions incorporate additional moments easily, and are satisfactory if atomic heterogeneity is not great, and often if the multiplicity of the space group is high, even if there is marked heterogeneity.

References and Footnotes

1. This is strictly true for the intensity distributions. There is a small systematic effect on the structure-factor distributions. [See (2), paragraphs 3.3 and 3.4, for discussion and references.]
2. A.J.C. Wilson, *Acta Crystallogr.* A *35,* 122-130 (1979).
3. A.J.C. Wilson, *Acta Crystallogr.* A *34,* 474-475 (1978).

4. A.J.C. Wilson, *Acta Crystallogr.* A *36,* 929-936 (1980).
5. A.J.C. Wilson, *Proc. Ind. Acad. Sci. Chem. Sci. 92,* 335-339 (1983).
6. A.J.C. Wilson, *Research 4,* 141-142 (1951).
7. A.J.C. Wilson, *Acta Crystallogr.* A *34,* 474-475 (1978).
6. A.J.C. Wilson, *Acta Crystallogr. 2,* 318-321 (1949).
7. A.J.C. Wilson, *Acta Crystallogr.* A *36,* 945-946 (1980).
8. H. Lipson and M. M. Woolfson, *Acta Crystallogr. 5,* 680-682 (1952).
9. A.J.C. Wilson, *Research 5,* 589-590 (1952).
10. D. Rogers and A.J.C. Wilson, *Acta Crystallogr. 6,* 439-449 (1953).
11. R. Srinivasan and S. Parthasarathy, *Some Statistical Applications in X-ray Crystallography,* chapter 3. Oxford: Pergamon Press (1976).
13. U. Shmueli, *Acta Crystallogr.* A *35,* 282-286 (1979).
14. U. Shmueli and A.J.C. Wilson, *Acta Crystallogr.* A *37,* 342-353 (1981).
15. U. Shmueli and A.J.C. Wilson, *Acta Crystallogr.,* A*39,* 225-233 (1983).
16. H. Cramér, *Mathematical Methods of Statistics.* Uppsala: Almqvist & Wiksells (1945).
17. A.J.C. Wilson, *Mathematical Theory of X-ray Powder Diffractometry.* Eindhoven: Centrex (1963).
18. It may be noted that the relations between the cumulants κ_k and κ_{k-l} are simply additive, and the coefficients of the orthogonal-polynomial expansions would be simpler if expressed in terms of the cumulants instead of the moments. The cumulant coefficients of the Edgeworth form of the Hermite expansion are given explicitly in (16), p. 171, and (19), p. 935, but those for Laguerre and subcentric expansions need to be worked out.
19. M. Abramowitz and I. A. Stegun, *Handbook of Mathematical Functions.* Washington: U.S. Govt Printing Office (1964).
20. G.H. Weiss and J.E. Kiefer, *J. Phys.* A*16,* 489-495.
21. U. Shmueli, G.H. Weiss, J.E. Kiefer and A.J.C. Wilson, *Acta Crystallogr.* A*40,* 651-660 (1984).
22. U. Shmueli, G. H. Weiss, J. E. Kiefer and A. J. C. Wilson, this volume, pp. 23-42.

Structure & Statistics in Crystallography, ISBN 0-940030-10-1,
Ed., A. J. C. Wilson F. R. S., Adenine Presss, ©Adenine Press 1985.

Precision and Accuracy in Structure Refinement by the Rietveld Method

E. Prince
Center for Materials Science
National Bureau of Standards
Gaithersburg, MD 20899, USA

Abstract

Whenever the values of a set of experimental observations can be predicted by a model containing adjustable parameters, the values of those parameters can be estimated by the method of least squares. In powder diffraction, if the shapes of individual, resolved peaks are known, the observations can be represented by a model of the form $y_i = b_i + \Sigma I_k P_{ik} + e_i$, where y_i is the intensity at point i in the pattern, b_i is the value of a background function at point i, I_k is the integrated intensity of the kth Bragg reflection, and e_i is a random error drawn from a population with zero mean. P_{ik} is a peak shape function, normalized so that the sum over a peak is equal to one, and the summation is performed over all Bragg peaks that can contribute to point i. Statistical methods may be used to test whether the fitted model is consistent with the data and, on the assumption that the model is the correct one, to estimate the standard deviations of the parameters. Standard deviations, however, are measures of precision rather than of accuracy. Various workers have attempted to assess the accuracy of the method, by defining a number greater than one by which the standard deviations may be multiplied, or by using alternative procedures, such as the separate estimation of integrated intensities or the use of non-diagonal weight matrices. All of these attempts are hampered by the absence, in the data, of any information concerning the correlations between systematic errors, and other sources of bias, and the parameters being estimated.

Introduction

The realization by Rietveld (1) that, if the underlying peak shape in a powder diffraction pattern is known, the information contained in individual members of sets of overlapping peaks is not completely lost unless they exactly coincide, has led to a revival over the past fifteen years of the use of powder diffraction as a tool for structure analysis. Although Rietveld's original computer program had a number of shortcomings, and most active workers today are using extensively modified or totally new versions, he is generally given credit for the concept of fitting the raw data in a powder pattern with a model involving adjustable parameters, and the technique has become known as the Rietveld method. It was noticed very soon, however, that the method frequently did not give estimates of some parameters, particularly thermal parameters, that were plausible within the apparent *precision* of the estimate, as computed by conventional methods in the least squares program. Because of this there have been a number of suggestions (2, 3) that the computation of standard deviations in the Rietveld method is inherently incorrect, and several attempts (3-7) to devise alternative procedures that give reasonable assessments of the *accuracy* of the method.

It is the purpose of this paper to show that, from the point of view of statistical analysis, the Rietveld method is a correct application of general principles; to try to identify the important problems that have led to the confusion and clarify the issues involved; and to discuss the current situation with regard to precision and accuracy in refinement using powder data. Because some of the sources of confusion have been rather basic misconceptions, we shall start with a discussion of the Gauss-Markov theorem and the conditions required for it to be applied. (These are frequently misstated in elementary texts on data fitting.) We shall then discuss the Rietveld method as an application of its basic principles. Third, we shall discuss some of the problems that arise in previous analyses. Finally, we shall consider to what extent the results can be relied upon and where there are remaining sources of potential bias that are inadequately understood.

The Gauss-Markov Theorem

Let us suppose that a set of observations, $\mathbf{y}$, represents measurements of phenomena that are described by a model, $M(\mathbf{x})$, which is a function of a set of parameters, $\mathbf{x}$, that are constants of nature, but whose values are unknown. The process of measurement introduces a random scatter into the observations, so that

$$y_i = M_i(\mathbf{x}) + e_i, \tag{1}$$

where e_i is a random variable drawn from a population with zero mean.

We shall use angle brackets to denote *expected values*, which are defined by

$$\langle f(z) \rangle = \int f(z) \Phi(z) \mathrm{d}z, \tag{2}$$

where $\Phi(z)$ is the probability density function for the distribution of a random variable, z. The mean of z is, by definition, $\langle z \rangle$. Rewriting equation (1) using this notation,

$$\langle y_i \rangle = \langle M_i(\mathbf{x}) + e_i \rangle = M_i(\mathbf{x}) + \langle e_i \rangle = M_i(\mathbf{x}), \tag{3}$$

since we have specified that $\langle e_i \rangle = 0$. The quantity

$$V_{ij} = \langle (z_i - \langle z_i \rangle)(z_j - \langle z_j \rangle) \rangle = \int (z_i - \langle z_i \rangle)(z_j - \langle z_j \rangle) \Phi_J(\mathbf{z}) \mathrm{d}\mathbf{z}, \tag{4}$$

where $\Phi_J(\mathbf{z})$ is the joint probability density function of the elements of a vector, $\mathbf{z}$, is the *covariance* of elements i and j. If $i = j$, it is the *variance*; the positive square

Precision & Accuracy

root of the variance is the *standard deviation*. If the elements of a vector, $\mathbf{x}$, are linear functions of the elements of another vector, $\mathbf{y}$, *i.e.*, if $\mathbf{x} = \mathbf{By}$, where $\mathbf{B}$ is some matrix, it is easily shown from the definitions that the means of the elements of $\mathbf{x}$ are related to the means of the elements of $\mathbf{y}$ by $\langle \mathbf{x} \rangle = \mathbf{B}\langle \mathbf{y}\rangle$, and that the variance-covariance matrix for $\mathbf{x}$ is related to the variance-covariance matrix for $\mathbf{y}$ by $\mathbf{V}_x = \mathbf{B V}_y \mathbf{B}^T$.

Let us consider now the case in which the model is linear, so that $\langle \mathbf{y} \rangle = \mathbf{Ax}$. If $\mathbf{A}$ has full column rank, the minimum of the quadratic form $(\mathbf{y} - \mathbf{Ax})^T \mathbf{W}(\mathbf{y} - \mathbf{Ax})$, where $\mathbf{W}$ is a positive definite matrix, is found at $\hat{\mathbf{x}} = \mathbf{Hy}$, where $\mathbf{H} = (\mathbf{A}^T\mathbf{WA})^{-1}\mathbf{A}^T\mathbf{W}$. The vector, $\hat{\mathbf{x}}$, is an *estimate* of $\mathbf{x}$. Because $\langle \hat{\mathbf{x}} \rangle = \langle \mathbf{Hy} \rangle = \mathbf{H}\langle \mathbf{y}\rangle = \mathbf{HAx} = (\mathbf{A}^T\mathbf{WA})^{-1}\mathbf{A}^T\mathbf{WAx} = \mathbf{x}$, $\hat{\mathbf{x}}$ is an *unbiased estimate* of $\mathbf{x}$ independent of $\mathbf{W}$, provided only that $\mathbf{W}$ is positive definite. The variance-covariance matrix for $\hat{\mathbf{x}}$ is $\mathbf{V}_x = \mathbf{H V}_y \mathbf{H}^T$. Now consider the matrix $\mathbf{G} = (\mathbf{A}^T\mathbf{V}_y^{-1}\mathbf{A})^{-1}\mathbf{A}^T\mathbf{V}_y^{-1}$ and the difference matrix $\mathbf{D} = \mathbf{H V}_y \mathbf{H}^T - \mathbf{G V}_y \mathbf{G}^T$. $\mathbf{Gy}$ is the particular unbiased estimate of $\mathbf{x}$ for which the weight matrix is $\mathbf{W} = \mathbf{V}_y^{-1}$. Since $\mathbf{V}_y$ is positive definite, there exists a non-singular, lower triangular matrix, $\mathbf{L}$, with positive diagonal elements such that $\mathbf{LL}^T = \mathbf{V}_y$. (Proofs that $\mathbf{L}$, called the *Cholesky decomposition*, exists for any positive definite matrix can be found in most texts on linear algebra.) It can be shown by straightforward matrix algebra that $\mathbf{D} = [(\mathbf{H} - \mathbf{G})\mathbf{L}][(\mathbf{H} - \mathbf{G})\mathbf{L}]^T$. The diagonal elements of $\mathbf{D}$ are the sums of the squares of the elements of rows of $(\mathbf{H} - \mathbf{G})\mathbf{L}$, and must therefore be greater than or equal to zero. It follows that the diagonal elements of $\mathbf{V}_x$, the variances of the estimated elements of $\mathbf{x}$, are minimum if $\mathbf{W} = \mathbf{V}_y^{-1}$. $\mathbf{V}_x$ is then given by

$$\mathbf{V}_x = (\mathbf{A}^T\mathbf{V}_y^{-1}\mathbf{A})^{-1}\mathbf{A}^T\mathbf{V}_y^{-1}\mathbf{V}_y\mathbf{V}_y^{-1}\mathbf{A}(\mathbf{A}^T\mathbf{V}_y^{-1}\mathbf{A})^{-1} = (\mathbf{A}^T\mathbf{V}_y^{-1}\mathbf{A})^{-1}. \qquad (5)$$

The previus paragraph is a proof of one of the important results of the Gauss-Markov theorem; it may be stated as follows: If $\langle \mathbf{y}\rangle = \mathbf{Ax}$, and the elements of $\mathbf{y}$ have a joint probability distribution describable by a density function with a variance-covariance matrix $\mathbf{V}_y$, then $\hat{\mathbf{x}} = [(\mathbf{A}^T\mathbf{V}_y^{-1}\mathbf{A})^{-1}\mathbf{A}^T\mathbf{V}_y^{-1}]\mathbf{y}$ is the unbiased linear estimate of $\mathbf{x}$ for which the variance of the estimate of every element of $\mathbf{x}$ is minimum. It should be noted that the proof does not require that the errors in the observations be independently distributed. It does require that they have a variance-covariance matrix that is positive definite and also that the model *correctly* represents the quantity that is being observed.

The discussion thus far applies only to linear models. We can apply it to non-linear models by assuming that, for some $\mathbf{x}'$ sufficiently close to the true $\mathbf{x}$, the linear function $M_i'(\mathbf{x}) = M_i(\mathbf{x}') + \mathbf{a}_i(\mathbf{x} - \mathbf{x}')$, where $\mathbf{a}_i$ is the ith row of the matrix whose elements are $A_{ij} = M_i(\mathbf{x})/\partial x_j$, and the partial derivatives are evaluated at $\mathbf{x} = \mathbf{x}'$, is a good approximation to $M_i(\mathbf{x})$. The estimate of $\mathbf{x}$ is then $(\hat{\mathbf{x}} - \mathbf{x}') = \mathbf{G}[\mathbf{y} - \mathbf{M}(\mathbf{x}')]$.

The Rietveld Model

If the shape of a powder diffraction peak can be represented by a function of 2θ, for fixed wavelength diffractometry, or time of flight, in the pulsed neutron source case, the observed diffraction pattern can be represented by a model of the form

$$y_i = b_i + \sum_{k=k_1}^{k_2} I_k P_{ik} + e_i, \qquad (6)$$

where y_i is the observed intensity at position i, b_i is the value of a background function, I_k is the integrated intensity of the kth Bragg reflection, and P_{ik} is a peak shape function, normalized so that the sum over a peak is one. The integrated intensity has the form

$$I_k = SL_k p_k m_k F_k^2, \qquad (7)$$

where S is a scale factor, L_k and p_k are the Lorentz and polarization factors, m is the multiplicity, and F_k is the structure amplitude. The actual observation is a count of neutrons or x-ray photons, which is subject to a random scatter having a Poisson distribution. The usual weight matrix is diagonal, with $W_{ii} = 1/y_i$. The question of the ideal weight matrix is discussed in detail below.

Lack of Fit

We have seen that the Gauss-Markov theorem is based on the hypothesis that the model correctly predicts the expected values of the observations. It is not necessary that the observations be statistically independent, but, if they are, the weight matrix is diagonal, with $W_{ii} = 1/\sigma_i^2$, where σ_i is the standard deviation of the ith observation. The quantity minimized is then a sum of terms of the form $\{[y_{io} - M_i(\mathbf{x})]/\sigma_i\}^2$. If $\mathbf{x}$ has its true value, then, from the definition of σ_i, $\langle\{[y_{io} - M_i(\mathbf{x})]/\sigma_i\}^2\rangle = 1$, from which it follows that the expected value of a sum of n terms is n. For each element of $\mathbf{x}$ that is estimated, the expected value of the sum is reduced by one, so that if the model has p parameters, $\langle \Sigma\{[y_{io} - M_i(\hat{\mathbf{x}})]/\sigma_i\}^2\rangle = n - p$. If the model *is* correct, then, the observed value of this sum should be close to $n - p$, and any higher value is an indication of lack of fit due to an inadequate model.

It is a common practice to take higher values of this sum as indications not of lack of fit but of systematic underestimates of the variances of the measurements. All elements of the variance-covariance matrix are then multiplied by a number greater than one in order to make the results consistent. In powder patterns of well crystallized materials, however, it is commonly observed that isolated peaks due to single resolved reflections can be fitted with peak shape functions of the assumed shape well within the statistically expected criteria. Each data point on one of these peaks is therefore an independent estimate of its integrated intensity, with the least squares fit giving a weighted mean. A systematic difference between the calculated profile and the independent fit to the observed profile must therefore be taken as being due to lack of fit, and cannot be treated as if the experimental data are less precise than they really are. The effect that lack of fit has on parameter estimates depends critically on the correlations between the individual refined parameters and whatever others are missing from the model. Two extreme cases are shown in fig. 1, in which contours of the sum of squares function are plotted for a refined parameter, plotted horizontally, and a missing parameter, plotted vertically. In one case the parameters are orthogonal to one another, and the lack of fit has no effect on either the refined parameter or its standard deviation. In the other case there is a strong correlation, leading to a bias in both the parameter and its standard deviation.

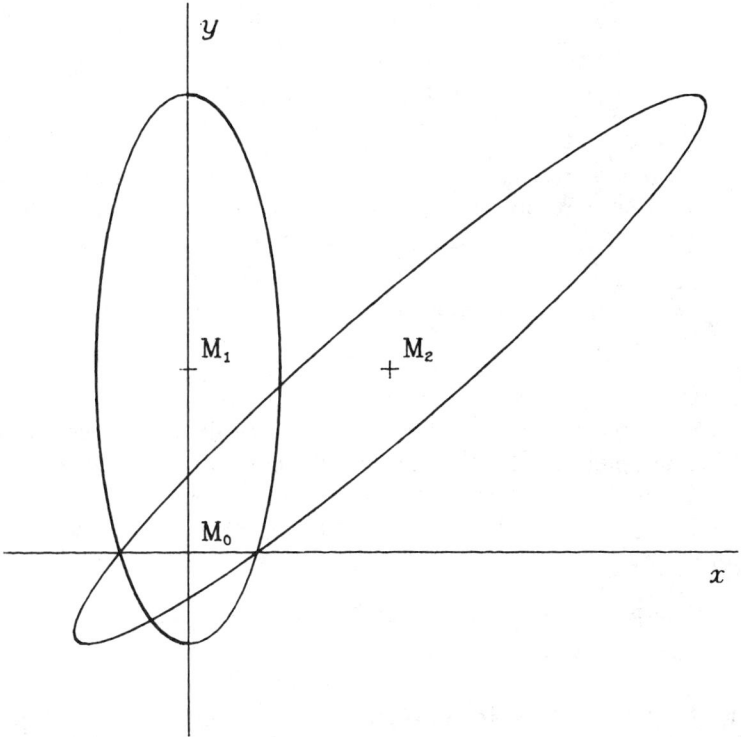

Figure 1. Contours of joint distributions in two lack of fit situations. The minimum for the incomplete model is at M_0. If the true minimum, when all parameters are included, is at M_1, the parameters plotted along x and y are uncorrelated, and the estimate of the x parameter is unbiased. If the true minimum is at M_2, the estimates of both the value and the standard deviation of the x parameter are biased.

Integrated Intensity Methods

The Bragg contribution to the powder pattern is the product of two factors, an integrated intensity, I_k, and a peak shape function P_{ik}. The integrated intensity is a function of atomic position and thermal parameters, whereas the peak shape is a function of cell constants and parameters describing the peak width as a function of 2θ or of time of flight. Because these sets of parameters are non-intersecting, and because the shape function is observed to be much less subject to theoretical uncertainties than the crystallographic model, Cooper (3) has suggested that the two types of parameter should be estimated separately in a two stage procedure in which the integrated intensities are first derived from the raw data and then used to determine the crystallographic parameters. Mathematically, this procedure involves partitioning the design matrix, $\mathbf{A}$, by columns into two matrices, $\mathbf{X}$ for the crystallographic parameters and $\mathbf{R}$ for all the rest. $\mathbf{X}$ is factored into two matrices, $\mathbf{S}$, with elements $S_{ik} = (\partial y_i/\partial I_k)/\sigma_i$ and $\mathbf{T}$, with elements $T_{kj} = \partial I_k/\partial x_j$. The normal equations matrix for the first stage of refinement is then

$$\mathbf{C} = \begin{bmatrix} \mathbf{S}^T\mathbf{S} & \mathbf{S}^T\mathbf{R} \\ \mathbf{R}^T\mathbf{S} & \mathbf{R}^T\mathbf{R} \end{bmatrix} \tag{8}$$

If, *and only if*, this matrix is non-singular, a variance-covariance matrix for the integrated intensities and the shape parameters can be written in the form

$$\mathbf{C}^{-1} = \begin{bmatrix} \mathbf{V}_s & \mathbf{V}_c \\ \mathbf{V}_c^T & \mathbf{V}_R \end{bmatrix}, \tag{9}$$

where $\mathbf{V}_s$ and $\mathbf{V}_R$ are variance-covariance matrices for the integrated intensities and the rest of the parameters, while $\mathbf{V}_c$ contains the correlations between the two sets. In accordance with the Gauss-Markov theorem, the weight matrix for the second stage should be $\mathbf{V}_s^{-1}$, and we therefore need an expression for $\mathbf{V}_s$ in terms of $\mathbf{S}$ and $\mathbf{R}$. Consider the matrix

$$\mathbf{Q} = \begin{bmatrix} [\mathbf{S}^T\mathbf{S} - \mathbf{S}^T\mathbf{R}(\mathbf{R}^T\mathbf{R})^{-1}\mathbf{R}^T\mathbf{S}]^{-1} & -[\mathbf{S}^T\mathbf{S} - \mathbf{S}^T\mathbf{R}(\mathbf{R}^T\mathbf{R})^{-1}\mathbf{R}^T\mathbf{S}]^{-1}\mathbf{S}^T\mathbf{R}(\mathbf{R}^T\mathbf{R})^{-1} \\ -[\mathbf{R}^T\mathbf{R} - \mathbf{R}^T\mathbf{S}(\mathbf{S}^T\mathbf{S})^{-1}\mathbf{S}^T\mathbf{R}]^{-1}\mathbf{R}^T\mathbf{S}(\mathbf{S}^T\mathbf{S})^{-1} & [\mathbf{R}^T\mathbf{R} - \mathbf{R}^T\mathbf{S}(\mathbf{S}^T\mathbf{S})^{-1}\mathbf{S}^T\mathbf{R}]^{-1} \end{bmatrix} \tag{10}$$

It is readily verified that $\mathbf{QC} = \mathbf{CQ}^T = \mathbf{I}$, so that, although $\mathbf{Q}$ does not appear to be symmetric, it is in fact a representation of $\mathbf{C}^{-1}$. The proper normal equations matrix for the second stage of an integrated intensity procedure is therefore $\mathbf{T}^T[\mathbf{S}^T\mathbf{S} - \mathbf{S}^T\mathbf{R}(\mathbf{R}^T\mathbf{R})^{-1}\mathbf{R}^T\mathbf{S}]\mathbf{T}$ or $\mathbf{X}^T\mathbf{X} - \mathbf{X}^T\mathbf{R}(\mathbf{R}^T\mathbf{R})^{-1}\mathbf{R}^T\mathbf{X}$, and the variance-covariance matrix for the crystallographic parameters is the inverse of this matrix. The normal equations matrix for the Rietveld method can be written

$$\mathbf{B} = \begin{bmatrix} \mathbf{X}^T\mathbf{X} & \mathbf{X}^T\mathbf{R} \\ \mathbf{R}^T\mathbf{X} & \mathbf{R}^T\mathbf{R} \end{bmatrix} \tag{11}$$

which has the same form as $\mathbf{C}$ in equation (8). By analogy with equation (10), therefore, the variance-covariance matrix for the crystallographic parameters in

Precision & Accuracy

the Rietveld method is $\mathbf{V}_x = [\mathbf{X}^T\mathbf{X} - \mathbf{X}^T\mathbf{R}(\mathbf{R}^T\mathbf{R})^{-1}\mathbf{R}^T\mathbf{X}]^{-1}$, which is identical to that for the integrated intensity method.

Thus, the Rietveld method is mathematically identical to properly formulated integrated intensity methods provided the model is correct and all of the matrices that need to be inverted during the calculation are non-singular. However, if the peak shape function is simultaneously non-zero for two or more reflections, $\mathbf{S}^T\mathbf{S}$ will have non-zero off-diagonal elements, and the errors in their integrated intensities will be correlated. If two reflections exactly coincide, the normal equations matrix will be singular, and it will be extremely ill conditioned if the reflections almost coincide. The much smaller matrix, $\mathbf{X}^T\mathbf{X}$, may still be non-singular provided there is different information in different parts of a composite peak. The Rietveld method can in fact be viewed as a method for estimating integrated intensities under the constraint that they correspond to a crystal structure.

Observations and Residuals

Some confusion has arisen because of uncertainty over what is meant by statistically independent observations. We have seen that, for some set of trial parameters, $\mathbf{x}'$, close to the true $\mathbf{x}$, the best estimate of the parameters is $\hat{\mathbf{x}} = \mathbf{x}' + \mathbf{G}[\mathbf{y} - \mathbf{M}(\mathbf{x}')]$. The best predicted values of the observations are $\hat{\mathbf{y}} = \mathbf{M}(\mathbf{x}') + \mathbf{AG}[\mathbf{y} - \mathbf{M}(\mathbf{x}')]$, and the *residuals*, $\mathbf{y} - \hat{\mathbf{y}}$, are given by $\mathbf{y} - \mathbf{M}(\mathbf{x}') - \mathbf{AG}[\mathbf{y} - \mathbf{M}(\mathbf{x}')]$, which can be written $[\mathbf{I} - \mathbf{AG}][\mathbf{y} - \mathbf{M}(\mathbf{x}')]$. The vector of residuals is therefore a linear function of the observations, and it has a variance-covariance matrix given by $\mathbf{V}_R = [\mathbf{I} - \mathbf{AG}]\mathbf{V}_y[\mathbf{I} - \mathbf{AG}]^T = \mathbf{V}_y - \mathbf{A}(\mathbf{A}^T\mathbf{V}_y^{-1}\mathbf{A})^{-1}\mathbf{A}^T$. Even if $\mathbf{V}_y$ is diagonal, $\mathbf{V}_R$ will not be, and the errors in the residuals are correlated. Now $\mathbf{x}'$ is an arbitrary origin, in parameter space, chosen so that the linear approximation will be adequate, and $[\mathbf{y} - \mathbf{M}(\mathbf{x}')]$ is a vector of observed values *relative to* the values predicted for $\mathbf{x} = \mathbf{x}'$. The usual procedure is to choose $\mathbf{x}'$ by an iterative numerical algorithm that finds the minimum of the sum of squares. If this algorithm goes to full convergence, the parameter shift, $\hat{\mathbf{x}} - \mathbf{x}'$, becomes null, and $\hat{\mathbf{y}} = \mathbf{M}(\mathbf{x}')$, so that the vector of observations and the vector of residuals are numerically identical. Because $\mathbf{x}'$ is chosen by a least squares procedure it is easy to make the error of supposing it to be an estimate itself rather than simply the origin with respect to which $\hat{\mathbf{x}}$ is estimated in the very last cycle of refinement, and to conclude that the observations are correlated, as the numerically identical residuals are.

The matrix $\mathbf{V}_R$ can be shown to have rank $n - p$ and is therefore singular, so that its inverse cannot be used as a weight matrix. Various workers have observed, however, that the residuals of neighboring data points, particularly points within a Bragg peak in a powder pattern, tend to be particularly strongly correlated. Because of this, it is suggested that the weight matrix (8) include off-diagonal elements reflecting the correlations of residuals for nearby points, or use an integrated intensity method that gathers the points within a Bragg peak together with a common weight. However, as we have seen, the assumption (equation 1) that is implicit in least squares is that $y_i = M_i(\mathbf{x}) + e_i$, where e_i is drawn from a population with zero mean. No one value of e_i contains any information concerning the value of any other e_j. It is in this sense that the observations are independent. The weight matrix is then diagonal, with $W_{ii} = 1/\sigma_i^2$, where σ_i^2 is the variance of the distribution of the population from which e_i is drawn. In counting neutrons or photons, provided electronic dead time effects can be neglected, the count has the Poisson distribution

$$P(k) = \lambda^k \exp(-\lambda)/k!. \qquad (12)$$

This distribution has mean λ and variance λ, so $\lambda_i = M_i(\mathbf{x})$ and $W_{ii} = 1/M_i(\mathbf{x})$. However, because $\mathbf{x}$ is unknown initially, $M_i(\mathbf{x})$ is also unknown, but the observed value, y_i, is an estimate of λ, and $W_{ii} = 1/y_i$ is a good approximation that can be used for the initial weight matrix. If y_i is small, this introduces a slight bias, which can be corrected by replacing the observed value by the calculated value after an initial estimate, but, even for moderate observed intensities (>100) this bias is entirely negligible in comparison with other sources of uncertainty.

Precision and Accuracy

We have seen (equation 5) that, if the expected values of the observations are correctly represented by the model and if the joint probability density function for the observations has a variance-covariance matrix $\mathbf{V}_y$, then $\mathbf{V}_x = (\mathbf{A}^T\mathbf{V}_y^{-1}\mathbf{A})^{-1}$ gives the variance-covariance matrix for the parameter estimates, $\hat{\mathbf{x}}$, and the square roots of the diagonal elements of $\mathbf{V}_x$ are the standard deviations of the elements of $\hat{\mathbf{x}}$. These standard deviations are measures of precision, that is measures of *lower limits* on the *widths* of the marginal probability density functions for the parameter estimates. Of course, what the experimenter wants to know is, "How accurate is the estimate? How closely does it represent nature?" Experience with the Rietveld method indicates that there is wide variation in the relationships between the indicated precision and the plausible accuracy of different kinds of parameter. Thus, the precision of refined cell constants is often extremely high, and, provided the instrument has been properly calibrated, the indications are that accuracy is also very good. On the other hand, the parameters describing the width of a peak as a function of 2θ, the *UVW* of Caglioti, Paoletti and Ricci (9), are very imprecisely determined, and often bear a rather poor resemblance to the values that are calculated from the known instrumental parameters. Fortunately, these parameters tend to be correlated mostly among themselves, and do not contribute significantly to uncertainties in other parameters.

The structural parameters, atomic positions and thermal motion tensor elements, have an intermediate status. In cases where the same compounds have been studied by both the Rietveld method and single crystal methods, or where interatomic distances and bond angles can be compared with usual values, the results are often reasonable, relative to the indicated level of precision. The situation with respect to thermal parameters is much less satisfactory, with many seriously implausible values having appeared in the literature. In situations where large anisotropies can be expected, the parameters generally indicate large amplitudes in the proper directions, but tensors that are not positive definite are common. Hewat (10) has shown that the effects of absorption and multiple scattering have 2θ dependences that mimic temperature factors, and failure to allow for these will introduce a negative bias into all thermal parameters. Sabine (11) has observed that the parameters can be very sensitive to particle size, which may be attributable to an extinction effect.

A number of workers (4,5) have attempted to define factors greater than one by which standard deviations can be multiplied in order to translate precision into accuracy. The problem with this approach is that the data contain no information concerning the interactions between systematic errors, or model inadequacies, and the estimated parameters. While statistical tests can give an indication of whether the model is consistent with the data, they cannot rule out the existence of other models that are equally consistent with the data, nor can they, if they indicate that the model is not consistent with the data, tell why. Conversely, if enough is known about an experimental effect to assess its influence on the uncertainties of parameter estimates, the information should be incorporated in the model, thereby correcting for the effect.

References and Footnotes

1. H. M. Rietveld, *J. Appl. Cryst.*, *2*, 65 (1969).
2. M. Sakata and M. J. Cooper, *J. Appl. Cryst.*, *12*, 554 (1979).
3. M. J. Cooper, *Acta Cryst.*, *A38*, 264 (1982).
4. G. S. Pawley, *J. Appl. Cryst.*, *13*, 630 (1980).
5. H. G. Scott, *J. Appl. Cryst.*, *16*, 159 (1983).
6. E. Baharie and G. S. Pawley, *J. Appl. Cryst.*, *16*, 404 (1983).
7. M. J. Cooper, *Z. Krist. 164*, 157 (1983).
8. A. Albinati and B. T. M. Willis, *J. Appl. Cryst.*, *15*, 361 (1982).
9. G. Caglioti, A. Paoletti and F. P. Ricci, *Nucl. Instrum.*, *3*, 223 (1958).
10. A. W. Hewat, *Acta Cryst.*, *A35*, 248 (1979).
11. T. M. Sabine, Personal communication, to be published (1983).

Structure & Statistics in Crystallography, ISBN 0-940030-10-1,
Ed., A. J. C. Wilson F. R. S., Adenine Press, ©Adenine Press 1985

Discussion of the Paper by E. Prince

J.S. Rollett
Oxford University Computing Laboratory, 8-11 Keble Road
Oxford OX1 3QD, England

The measurements of intensity that are used in the refinement of crystal-structure parameters are subject to errors, which include:

(a) the random errors of quantum counting, with well understood statistics;
(b) errors due to absorption and extinction; and
(c) other errors such as those caused by thermal diffuse scattering.

The analysis is also falsified by imperfections in the model used to calculate intensities for comparison with those observed.

In principle all these errors should be removed or reduced to negligible levels. This is not done, either because it is too costly (for example to count quanta for longer times), or because the analysis is difficult or even impossible, or because there are insufficient data of adequate quality to determine the number of parameters needed (as in the case of elaborate models of vibration). In practice it is usually necessary to work with errors of many kinds, and to accept that it is very difficult to make reasonable estimates of the variances associated with all of them.

The distinction between errors of observation and those of the calculated intensities is not absolute. An error in correcting absorption belongs to the observed intensity if this is regarded as the intensity for an ideally small crystal. The same error belongs to the calculated intensity if it is regarded as the intensity of the actual crystal used. The latter choice is preferable in some ways, but is computationally expensive and inconvenient if the analysis is based on data from many crystals.

The distinction between random and systmatic error is not absolute. An absorption error is systematic (that is to say, not variable) if we consider merely repetition of measurements on a particular crystal specimen. If, however, we consider a population of measurements on a variety of comparable specimens, then the absorption effects, and the errors remaining after their correction, vary and can be considered to have a probability distribution. It is then just as reasonable to estimate standard deviations for structural parameters arising from this variation as it is to do the same thing for quantum counting.

Errors of the vibration model can be considered similarly. We can, at one extreme, say that we are analysing the vibrating (and possibly disordered) molecules in a particular specimen of crystal at a certain temperature. Then all vibration-model errors belong to the calculated intensities. Another extreme position is that we wish to refer everything to a crystal structure at rest, and that all imperfectly corrected deviations from that situation are errors to be assigned to the observed intensities. If we consider a population of sets of observed intensities for crystals in various conditions, and of sets of calculated intensities based on various reasonable models, then the differences between the observed and calculated intensities vary, can be considered to have a probability distribution, and produce variations in the structural parameters that can be described by standard deviations.

The distinction between errors of observation and those of calculation is not important for discussion of errors of parameters. The same size of error in either of the two will produce just as big a change in the results of the analysis. The distinction between random and systematic errors is not important either. It depends on the ensemble of experiments that we choose to imagine.

We need to consider what description of the errors of structural parameters will usually be the most useful. In general, functions of these parameters are compared with values from other analyses of crystal and molecular structure to study the effects of variations in chemical environment upon bond lengths, angles, and so on. For this, the relevant standard-deviation estimates are those derived from the variance/covariance matrix that corresponds to the effects of all the errors that contribute to the differences between observation and calculation. The problem is how to find this variance/covariance matrix.

The method of least squares determines the structural parameters as linear combinations (or linearized combinations) of the observations. Consequently the variance/covariance matrix of the parameters can be determined, if the linearization is accurate, from the variance/covariance matrix of the errors in the differences between the observed and calculated intensities. If the weight matrix is the inverse of the variance matrix for the differences the calculation is simple, but it can be done even if that is not so.

This leaves us with the problem of estimating the variance/covariance matrix for the total errors of the intensities that we use. Experience shows that attempts to do this from first principles, by enumerating error types and estimating their sizes, commonly fail. A much safer procedure is that introduced by Cruickshank (1), which consists of using the statistics of the residuals to estimate the total errors. The residuals are biased by the fitting of parameters to the observations, but this bias is small in most cases and probably negligible in comparison with the errors inherent in any method of estimation, at least when there are many more observations than parameters.

We do need to have residuals that are not more likely to be of one sign than of the other, if the parameter values are not to be biased. Provided that known errors have been corrected as far as possible this is likely to be so. Weighting systems based on a small number of parameters commonly give weighted residuals of essentially constant variance and zero correlation. It is believed that under these conditions the method of least squares with weights based on the empirical statistics of the residuals gives estimates of total error that are realistic.

Reference

1. D.W.J. Cruickshank (1949). *Acta Crystallogr.* 2, 65.

Comments on the discussion of J. S. Rollett by E. Prince

The term "standard deviation" in statistical usage has a precisely defined meaning, namely the positive square root of the second moment of probability density function about its mean. An "estimated standard deviation" is a function of experimental observations that, under certain conditions, can be shown to have an expected value equal to the standard deviation of the appropriate density function. In least squares one of the conditions is that the mean of the density function appropriate to the observations is given by a model containing parameters that have correct (although unknown) values. In practical cases it is rarely, if ever, possible to prove that this condition is met. Statistical tests for goodness of fit can only indicate that the model is consistent with the observations.

If the weight matrix is the inverse of the variance-covariance matrix for the observations, the inverse of the normal equations matrix is an estimate of the variance-covariance matrix of the parameter estimates. This is not, however, simply a matter of computational convenience, as seems to be implied in Dr Rollet's discussion. Rather, it is with this weight matrix that the variances of the estimates of all parameters are minimized (the Gauss-Markov theorem). It is always possible, by counting for a long enough time, to reduce the effects of random errors that are expressed by standard deviations to the point where they make negligible contributions to the uncertainties in the values of the parameters.

Systematic errors, or model inadequacies, make contributions to the uncertainties that are not random, that is they introduce biases, and assessments of uncertainty that try to account for these should therefore not be referred to as standard deviations. Different systematic errors in repeated experiments may introduce an additional randomness to the uncertainty, but the bias is not removed. There is indeed an important distinction between the random errors of observation and systematic errors that are not properly included in the calculation. Although analysis of residuals may give information about the nature of the model inadequacy, its effect can only be assessed by correcting the model and observing the changes that are thereby made in the parameters.

Editorial Note

The significance of bias, random error, and systematic error is discussed, with some references to the literature, in the Introduction (pp. v-vi above). Several of the following papers contain relevant material; page references may be found in the Index, pp. 213-225 below.

Parameter Estimation by Entropy Maximization

Douglas M. Collins
Department of Chemistry
Texas A&M University
College Station, TX 77843

Abstract

The method of least squares has been a powerful means of fitting a model to a set of observations. Its success notwithstanding, least squares suffers a number of deficiencies which invite alternative methods.

Jaynes' extension of Shannon's entropy provides a framework of consistent inference for adjustment of prior proportions to reflect new information. With a fixed set of experimental observations as (normalized) prior proportions m, and calculated values taken to be (normalized) p,

$$H = -\sum_i p_i \ln[p_i/m_i]$$

is to be maximized with respect to model parameters. The formalism is appropriate to any set of positive-definite observations and its detail is indicated for crystallographic problems of two major types.

Introduction

Any positive-definite set of reference values may be normalized and said to give the probabilities of certain outcomes. Whether or not the outcomes correspond to a physical apparatus, in principle they characterize a potential communications device or channel of certain information capacity. Shannon [1] proposed as the measure of information capacity an entropy,

$$H = -\kappa \sum_i p_i \ln p_i, \qquad (1)$$

where κ is a scale factor and p is the probability of an outcome. Jaynes' consideration of prior distributions [2] led to a relative entropy

$$H = -\sum_i p_i \ln[p_i/m_i], \qquad (2)$$

where the prior distribution m is normalized as is p, and H is nonpositive. Relative entropy may be viewed operationally as a measure of the information content of the distortions which change m into p.

Deeper significance for equation (2) is found in work (3, 4, 5) which has foundations wholly other than information theory. Instead, starting from a requirement that new data be used in a procedure of consistent inference, it was found that correct adjustment of a prior distribution is uniquely characterised as constrained maximization of H given by equation (2). Of course this is in complete agreement with the idea of information theory that the least biased use of new data is the one which adds to m as little information as possible to get p which agrees with the available data.

Skilling (6) has pointed out that in a procedure of consistent inference it is not necessary to interpret p and m as probabilities, but it is sufficient that they be any proportions. It is in this spirit that the present work and two applications are outlined, with intensities from a diffraction experiment as the proportions. Both applications involve the adjustment of parameters by a criterion of consistent inference or entropy maximization. In one case the parameters are scale factors relating intensity subsets to a common scale, and in the other the parameters correspond to the customary description of atomic crystal structures.

Outline

Problems of parameter estimation by entropy maximization are readily identified as involving either direct or indirect observations. The distinction corresponds to choice of the function on which entropy is maximized. In an earlier application in which electron density itself is the object of interest, entropy is maximized on the density subject to structure-factor constraints (7). Because the function of maximum entropy and the data are set in different spaces, this is an example of indirect observation and it is characterized by a variable prior proportion.

In contrast, problems involving direct observation are characterized by fixed prior proportions. In the present applications it is assumed that after correction by Lorentz, polarization, and other appropriate factors, the observed intensitites I are proportional to $|F|^2$, the structure-factor squared moduli. However, because the functions used in entropy expressions are normalized, the overall scale of $|F|^2$ is immaterial and any scale difference between $|F|^2$ and I is ignored. For entropy-maximization calculations, the fixed prior proportions may be taken as the corrected observed intensities to reflect the experimental background of the problem.

The problem at hand is to maximize

$$S = -\sum_h T'_h \ln[T'_h/\hat{T}'_h] \qquad (3)$$

Parameter Estimation by Entropy Maximization

by setting

$$\partial S/\partial p = 0,$$

where the sum over h includes all available observations, T formally represents intensity and is proportional to it, primed quantities are normalized after the pattern $T' = T/\Sigma T_h$, the circumflex denotes fixed experimental observations, and p is any parameter of the model from which T is calculated. The formal result and principal statement of this outline is

$$\sum_h \left\{ \frac{\partial T_h}{\partial p_i} - T'_h \left(\sum_H \frac{\partial T_h}{\partial p_i} \right) \right\} \ln[T'_h/\hat{T}'_h] = 0 \tag{5}$$

after substantial but straightforward algebraic computation.

The first application involves parameter estimation for customary description of atomic crystal structures. In this case the following identifications are made using A and B as the real and imaginary parts of F:

$$\hat{T} = |\hat{F}|^2 ;$$
$$T = |F + \sum_i (\partial F/\partial p_i)\, \delta p_i + ...|^2 ;$$
$$A_{hi}^{(1)} = \partial A_h/\partial p_i,\ A_{hi}^{(2)} = \partial^2 A_h/\partial p_i^2 ;$$
$$a_{ij} = \sum_h w_h [(A_{hi}^{(1)} A_{hj}^{(1)} + B_{hi}^{(1)} B_{hj}^{(1)}) + \delta_{ij}(A_h A_{hi}^{(2)} + B_h B_{hi}^{(2)})] ; \tag{6}$$
$$v_i = -\sum_h w_h [(A_h A_{hi}^{(1)} + B_h B_{hi}^{(1)}) - |F_h|^{2\prime} X_i] ;$$
$$X_i = \sum_h (A_h A_{hi}^{(1)} + B_h B_{hi}^{(1)}) ;$$
$$w = \ln[|F|^{2\prime}/|\hat{F}|^{2\prime}] .$$

After all necessary evaluations with parameters at current values, parameter shifts are found from solution of

$$a \delta p = v, \tag{7}$$

a matrix equation obtained by rearrangment of equation (5). It is not clear how observational weights might fit into this scheme. It may be necessary to consider using no weights beyond the binary choice of using or disregarding any particular observation. Standard deviations for the parameters of the structure model can be estimated at any time by usual methods.

The second application involves estimation of scale factors to relate intensity subsets to a common scale. In this case the following identifications are made:

$$\hat{T} = \hat{I} ;$$
$$T = kI ; \tag{8}$$

where k_i, which puts I on the scale of observation set i, and the maximum-entropy mean intensity I are the parameters of the model. After rearrangement of equation (5), the parameters are iteratively determined, as though independent, by solution of

$$\ln k_i = A - \sum_h I'_h \ln[I_h/\hat{I}_{hi}]$$
$$\ln I_h = A - \sum_i k'_i \ln[k_i/\hat{I}_{hi}] \; ; \qquad (9)$$
$$A = \sum_{h,i} (k_i I_h)' \ln[k_i I_h/\hat{I}_{hi}] \; .$$

This application is completely successful in practice as described elsewhere (8).

A remarkable feature of these results is that entropy maximization completely frees both applications from overall scale determination. This can be seen by inspection of equations (7), (8). For if after some iterations the scale of observation is arbitrarily changed, the solutions of the equations are unaffected.

Independence of overall scale provides an answer to the sometimes disabling problem of correlation between scale factor and thermal parameters in standard least-squares refinement. Moreover, in comparison with standard refinement, equations (7) appear to provide substantial decoupling of all parameter shifts. Although practical evaluation of this point is needed, among important possible consequences is an extensive new application of structure refinement through radical diagonalization techniques which do not work in least-squares calculations.

Entropy maximization has commanded attention for a variety of reasons. One important reason is that it is effective in application and has been a fruitful means to solutions of difficult problems in many fields. But now, the most telling of all is that if and only if calculations satisfy a principle of maximum entropy is the consistent use of all data ensured (3, 4, 5).

Acknowledgment

This work has been supported in part by the Robert A. Welch Foundation through grant A-742 and by the Research Corporation through a Cottrell Research Grant.

References and Footnotes

1. Shannon, C.E. & Weaver, W. *The Mathematical Theory of Communication*, University of Illinois Press, Urbana (1949).
2. Jaynes, E.T. *IEEE Trans. SCC-4*, 227-241 (1968).
3. Shore, J.E. & Johnson, R.W. *IEEE Trans. IT-26*, 26-37 (1980).
4. Shore, J.E. & Johnson, R.W. *IEEE Trans. IT-29*, 942-943 (1983).
5. Tikochinsky, Y., Tishby, N.Z. & Levine, R.D. *Phys. Rev. Lett. 52*, 1357-1360 (1984).
6. Skilling, J. *Nature (London) 309*, 748-749 (1984).
7. Collins, D.M. *Nature (London) 298*, 49-51 (1982).
8. Collins, D.M. *Acta Cryst. A40*, 705-708 (1984).

Structure & Statistics in Crystallography, ISBN 0-940030-10-1,
Ed., A. J. C. Wilson F. R. S., Adenine Presss, ©Adenine Press 1985.

Information and Crystal-Structure Estimation

Stephen W. Wilkins
Laboratory of Molecular Biophysics,
Department of Zoology, University of Oxford
South Parks Road, Oxford. OX1 3PS, England
and
C.S.I.R.O., Division of Chemical Physics,
P.O. Box 160, Clayton 3168, Victoria, Australia*

Stig Steenstrup
Physics Laboratory II, H.C. Orsted Institute,
Universitetsparken 5, 2100 Copenhagen, Denmark

and

Joseph N. Varghese
C.S.I.R.O., Division of Protein Chemistry,
Royal Parade, Parkville 3052, Victoria, Australia

Abstract

The conceptual foundations of a general information-theoretic based approach to X-ray structure estimation are re-examined with a view to clarifying some of the subtleties inherent in the approach and to enhancing the scope of the method. More particularly, general reasons for choosing the minimum of the Shannon-Kullback measure for information as the criterion for inference are discussed and it is shown that the minimum-information (or maximum-entropy) principle enters the present treatment of the structure-estimation problem in at least two quite separate ways, and that three formally similar but conceptually quite different expressions for relative information appear at different points in the theory. One of these is the general Shannon-Kullback expression, while the second is a derived form pertaining only under the restrictive assumptions of the present stochastic model for allowed structures, and the third is a measure of the additional information involved in accepting a fluctuation relative to an arbitrary mean structure.

The matter of choosing appropriate constraints for incorporating the available diffraction data and some of the prior knowledge into the minimum-information estimate for the structure is also discussed, and a particularly simple method for incorporating one's degree of belief in the prior structure is outlined.

*permanent address

1. Introduction

From their beginnings some 70 years ago, workers in X-ray diffraction have evolved a wide variety of procedures with which the structures of many small to medium-sized molecules have been elucidated. More recently, these procedures have been used to determine the structures of much larger molecules such as proteins and viruses, and to obtain structural information on biological entities such as membranes, nerves and muscle. With extension into the area of increased complexity and size, conventional methods have been found to become increasingly time consuming and unreliable. There is therefore a need to seek new, more efficient and less subjective approaches to structure estimation.

In some recent work, (1,2,3) we have outlined a very general approach to X-ray structure estimation which is capable of simultaneously encompassing conventional approaches to this problem in a highly self-consistent fashion and of incorporating entirely new sorts of information. The methods is based on the Shannon (4) and (more generally), Kullback (5,6) measures for information, and the minimum-information (or maximum-entropy) principle of Jaynes (7).

In the present work we re-examine the conceptual foundations of our earlier work with a view to clarifying some of the subtleties inherent in the approach and to enhancing the scope of the method. The outline of the work is as follows. Section 2 is concerned with the question of a measure for relative information and the minimum-information (MI) method of inference based on this quantity. Section 3 deals with the application of the MI principle to the X-ray structure-estimation problem *via* the assumption of a simple stochastic model for the assumed event space of allowed structures, while Section 4 deals with the incorporation of the data and prior knowledge into the MI estimate for the structure.

2. A Measure for Information and a Principle for Inference

From the work of Shannon (4) and Kullback (5,6) it is known that there exists a unique and consistent measure for the information content of a probability distribution, and that, for the case of the additional information contained in a distribution $P(x)$ relative to an initial distribution $P^\circ(x)$, it is given by

$$I[P;P^\circ] \equiv \sum_x P(X) \ln[P(x)/P^\circ(x)] \qquad (1)$$

where x is a random vector and we have assumed discrete probability distributions for simplicity. The functional I is known in different places under different names, e.g. cross-entropy, (8) directed divergence (5,6) and relative entropy (9).

However, the key property of (1) mentioned above, namely that it represents the additional amount of information involved in changing an initial distribution P° into a current distribution P, suggests to us that the term relative information is the most appropriate.

The functional (1) can be characterized axiomatically both in the discrete case (10,11) and in the continuous case (12), and shown to satisfy uniquely axioms which one might reasonably demand of a measure of relative information. The existence of such a measure lies at the basis of Jaynes (7,13) principle of maximum entropy (minimum information) for treating problems of inference, since one may sensibly assert that the least prejudiced choice for distribution which fits all available information is that which simultaneously minimizes (1) subject to a given prior. This principle for inference has some eminently desirable properties including the one that the minimum-information distribution is equal to the frequency distribution which can be realized in the greatest number of ways. (13) Nevertheless, there remain many people for whom arguments of Jaynes and followers as to the basis for this principle remain unconvincing. Recently Shore and Johnson (14) have inverted the argument and shown that reasonable axiomatic demands, of a consistency type, on an abstract inferential operator which combines an initial distribution $P^o(x)$ with mean-value constraints, leads precisely to the same distributions as would be obtained by minimizing I given by (1) subject to the same information. The axioms invoked by Shore and Johnson include uniqueness and the requirement that it should not matter whether one accounts for information about independent systems sequentially or simultaneously. Their arguments are extremely convincing and essentially show how the key elements of probability theory can provide a consistent model for inductive inference via the minimum-information principle.

3. A Simple Model for Feasible Structures and its Consequences

3.1 Choice of Event Space

In the case of the X-ray structure-determination problem, the quantity which we wish to estimate is the electron density $\rho(r)$, in the unit cell. For simplicity, let us restrict consideration to the one-dimensional case (although all results hold for arbitrary dimension) and let us work with the discrete and normalized density, $\rho = \{\rho_1,...,\rho_N\}$, corresponding to $\rho(r)$, where ρ_j is the normalized density for the jth pixel ($j = 1,...,N$) and leads to the unitary structure factor

$$U_k = |U_k|\exp\{i\phi_k\} = \langle \sum_{j=1}^{N} \rho_j \exp\{2\pi ijk/N\}\rangle$$
$$= \sum_j p_j \exp\{2\pi ijk/N\}, \quad (2)$$

where $p_j = \langle \rho_j \rangle$ and the expectation value is taken w.r.t. $P(x)$.

The random vector whose distribution we wish to describe is thus $x = \rho$ which also happens to have the characteristics of a probability distribution, since $\sum_{j}^{N} \rho_j = 1$.

In order to apply an information-theoretic approach to X-ray structure estimation, we first need to establish our allowed sample space or event space. Ideally, the sample space should be such as to incorporate as much of our prior knowledge and belief about the class of allowed structures, including concepts such as atomicity, stereochemistry, biochemistry, *etc*. However, these latter forms of information, and particularly those involving different types of chemical knowledge, are difficult to formulate in compact mathematical form. For this reason we take the following somewhat different approach (see also Refs. 1, 2, 3 and 15). We choose a sample space which is tied to a particularly simple model of the allowed structures, namely those capable of being generated by the independent distribution of n identical discrete units of structure over the N pixels with prior probability density q_j, in the manner of Bernoulli trials, and described by the occupation levels of the N pixels by $\boldsymbol{n} = (n_1, n_2, ..., n_N)$. The discrete units of structure may best be conceived of as being elements of resolution in ρ-space, rather than as specific entities such as atoms, amino acids, or electrons, although these are not perforce excluded. The principal virtue of this highly simplified event space is that it is obviously sufficiently general to incorporate all conceivable structures up to a certain resolution limit, provided n is large enough. On the other hand, its principal deficiency is that it does not incorporate as much of the available prior knowledge as one would like. Such information may, however, be incorporated into the problem by appropriate choice of prior distribution, $P(x)$, or the adoption of appropriate constraint relations.

3.2 Choice of Prior Distribution

The prior distribution, $P^{\circ}(x)$, expresses the probability with which we expect different structures to occur before we introduce the data and any new information. In the case of the simple "n ball" model outlined above (sometimes called the "monkey and the n balls model" (15), the prior probability of getting different occupation levels, $\boldsymbol{n} = \{n_1, ..., n_N\}$, is

$$P^{\circ}(\boldsymbol{n}) = n! \prod_{j=1}^{N} \frac{(q_j^{n_j})}{n_j!} \qquad (3)$$

and is a multinomial distribution. Clearly, a crystallographer given the same task of distributing n elements of structure over N pixels would be in no better position than a monkey for the first few "balls", but as the number of "balls" placed became large, the crystallographer would tend to "recognize" elements of structure and to develop beliefs as to where future placements of the elements might go, thus deviating from the simple notion of Bernoulli trials. Nonetheless, we adopt the simple form (3) with its associated assumptions and assume that all information on correlations between placements comes essentially from the data and any constraint relations (See Sec. 4).

3.3 A Derived Expression for Information in Terms of Mean Occupation levels of Pixels.

Instead of working with (1), which involves sums over macrostates, $\boldsymbol{n}$, of the whole

Information & Crystal-Structure Estimation

system, it is simpler to work with sums over mean occupation levels of microstates. More specifically, it is assumed that only mean levels of occupation numbers, $\langle n_j \rangle$, are known, viz.

$$\langle n_j \rangle / n = p_j = \langle \rho_j \rangle \qquad j = 1, ..., N \tag{4}$$

so that by a simple extension of some results given by Levine, (16) minimization of the information $I[P;P^o]$, given by (1), w.r.t. the N constraints (4) yields the following derived expression for the (structural) information in this case:

$$I^{MI1}[\boldsymbol{p};\boldsymbol{q}] = n \sum_{j=1}^{N} p_j \ln[p_j/q_j] \tag{5}$$

and leads to a minimum information (MI) distribution for $\boldsymbol{n}$ of multinomial form

$$P^{MI1}(\boldsymbol{n}) = n! \prod_{j=1}^{N} (p_j)^{n_j} / n_j! . \tag{6}$$

So far, the p_j simply represent arbitrary parameters having the properties of a probability distribution. However, since it is our overall intention to make information-theoretic estimates of crystal structures, it is manifest that all predictions which we make should be such as to fit our constraints or restraints (see Sec. 4) and minimize (1). By the sequential inference property of I discussed in Sec. 2, it matters not whether we go directly from (1) and the constraints or from (5) and the constraints, provided our constraints are functions of $\boldsymbol{p}$ (as holds in the cases which we shall consider in this work). The close formal similarity between (10) and (5) would have suggested this approach anyway, and in fact (5) has often been taken as the starting point for MI methods of image and structure processing. Thus, it is implicit in our approach that $\boldsymbol{p}$ is eventually to be chosen so as to fit any available data and to minimize (5). The resulting quantity we will denote by $\boldsymbol{p}^{MI2}$, since it represents a second stage application of the MI principle in the present context.

To summarise, the information function (5) is derived from (1) and embodies some very specific prior information and assumptions:

- I1: discrete positive integer $n_j = 0, 1, 2, ..., n$
- I2: prior probabilities for states $\boldsymbol{n}$ given by (3)
- I3: fixed total number of resolutional elements, i.e. $\sum_{j=1}^{N} n_j = n$.
- I4: constraints to be introduced later are functionals of $\boldsymbol{p}$ (not just $\boldsymbol{n}$) as exemplified by (2).

In the context of the crystal-structure estimation problem, these mathematical assumptions have the immediate consequence that:

A1: only those structures are considered where the total number of electrons exactly equals the total number (z) of electrons in the unit cell.

A2: only structures with positive discrete bounded density ρ are considered.

A3: further prior information and assumptions about the structure can and should be injected into p^{MI2} via q and appropriate constraints (see Sec. 4).

3.4 Fluctuations from the Mean Level of Occupation

The MI distribution (6) expresses the probability of a fluctuation ($n/n-p$) relative to an arbitrary mean level (structure), p. Where p is a minimum-information structure, p^{MI2}, w.r.t. some given information, then $p^{MI1}(n)$ expresses the probability of a fluctuation, ($n/n - p^{MI2}$) relative to the MI structure p^{MI2}.

As such it represents the best estimate for the possibility of that fluctuation based on the available information. Moreover, because the Fourier-transform relation between p and U_k is a unitary one, the probability of the fluctuation ($n/n - p^{MI2}$) is equal to the probability of the corresponding fluctuation in structure factors relative to the MI estimates for the structure factors corresponding to the same given information in reciprocal space. Recently Bricogne (17) has shown in detail how, under particular restricted assumptions on the form of the given information, such a probability relation (in the form given in (7) below) provides a most general path to the formulation of probabilistic direct methods without the limitations and approximations inherent in traditional approaches to direct methods.

It should be noted that (6) is an exact expression for the MI distribution of n within the present assumptions and model, and is in fact the conjugate distribution of $P^0(n)$ which is also multinomial (see *e.g.* p. 111 of Ref. 6).

If we now assume n to be large in (6) and invoke Stirling's approximation, we find that

$$-\ln P^{MI1}(n) \simeq I(n/n;\langle n\rangle/n) \equiv I(n/n;p)$$
$$= n \sum_{}^{N} (n_j/n)\ln[(n_j/n)/p_j] \qquad (7)$$

gives a measure of the additional information, $I(n/n;p)$, introduced into a map relative to the information of the mean map, $\langle n\rangle/n = p$. Within the asymptotic regime for n, it also follows that the variance-covariance matrix of the n_j/n is given by

$$\langle(n_j/n-p_j)(n_{j'}/n-p_{j'})\rangle = \left(n^2\frac{\partial^2 I(n/n;p)}{\partial n_{j'}\partial n_j}\right)^{-1}$$
$$\simeq \frac{1}{n}[\delta_{jj'}p_j - p_j p_{j'}] \qquad (8)$$

Information & Crystal-Structure Estimation

and tends to zero as $n \Rightarrow \infty$. This expression provides some insight into the role of n in the present formalism. In particular, when n has a clear-cut physical meaning, one can estimate the likely fluctuations in a map, or conversely, if one has an idea of the likely fluctuations in a map, then one can estimate n. Expressions (8) suggests that, within the present highly simplified model for structures, the value of n in practice is closely tied to both the quality and quantity of the data. From the derivation of (8) it may be seen that the $\delta_{jj'} p_j$ term on the r.h.s. of (8) arises solely from the positivity constraint on the $n_{j'}$ whilst the $p_{j'} p_j$ term arises solely from the normalization constraint on the n_j.

Fourier inversion of (8) yields a corresponding result in reciprocal space, namely

$$\langle (U_{k'}^{(n)} - U_{k'})^* (U_k^{(n)} - U_k) \rangle \simeq \frac{1}{n} [U_{k-k'} - U_k U_{-k'}] \tag{9}$$

with term-by-term correspondence with (8). In the present context, one can thus see how positivity of ρ in direct space immediately implies correlations in reciprocal space (see also Ref. 18).

3.5 Some Thoughts on the Meaning of n

Undoubtedly one of the most contentious points in the whole development of MI-based methods of crystal-structure estimation is the precise interpretation of n. Even though this matter has no practical consequences in so far as the determination of p^{MI2} w.r.t. some given information goes (since n merely rescales all the Lagrange multipliers in the problem), the meaning of n is important in calculating the probabilities of fluctuations from p^{MI2} via, say, (7). The form of MI-based predictions for the phases of structure invariants via such estimates for probabilities of fluctuations has been investigated recently by Bricogne (17), in which work n is identified with the number of atoms, n_{atoms}, *à la* traditional probabilistic direct methods. However, our earlier arguments lead us to the conclusion that n must depend on the quality and quantity of the data. To see that Bricogne's identification of n with n_{atoms} cannot be correct in general, let us consider the hypothetical case where the diffraction data for a system consist of a given number of structure factors (*i.e.* resolution). Then the reliabilities of traditional direct methods estimates for phases of invariants vary as $(1/n_{atoms})^{1/2}$ and higher powers of this quantity (see e.g. Ref. 19). Thus they individually tend to lose phase-determining power as n_{atoms} becomes large. On the other hand Bricogne's identification of n in (7) with n_{atoms} would lead one to the diametrically opposed conclusion that probabilities of structural fluctuations (and hence phases of invariants; *e.g.* see Ref. 19) tend to become sharper and sharper as n_{atoms} becomes large, and in fact, that the MI structure, p^{MI2}, is the only possible structure as $n_{atoms} \Rightarrow \infty$. This is clearly a *reductio ad absurdum,* since this is precisely the limit in which one knows least about the structure at atomic resolution, given that the number of structure factors (resolution) in the data remains fixed. By contrast our interpretation of n leads us to the interpretation that n is a function of the quantity and quality of the data, but not directly of n_{atoms}.

3.6 The Concept of Atomicity

While it is a physical fact that crystal structures consist of atoms and ions, often in bonded form, such knowledge has not been explicitly built into our model for the event space, *x*. In order to try to incorporate such information into the problem within the present general framework, one might proceed in two quite distinct ways. First, one might try to build such knowledge directly into the event space, say, *via* restrictions on the occupancies of pixels (McLachlan (20)) and also by imposing short-range correlations between occupancies of pixels. Alternatively, one might try to impose atomicity on the derived quantity *p* by introducing atomicity-type constraints (see *e.g.* Ref. 21). Neither approach has yet been developed to any significant extent within the information-theoretic framework, although to our minds, the latter path seems the easier to follow.

4. Crystal Structure Estimation via the Minimum-Information Principle

Thus far our development of an MI approach to structure estimation has been largely formal, especially as regards the nature of the information which p^{MI2} is required to satisfy. In this section we briefly touch upon some particularly useful cases of such information and assumptions. These cases are to be regarded as illustrative but by no means exhaustive, as to the avowed aim of any MI-based approach to structure estimation is to incorporate as much as possible of the available information and assumptions into a self-consistent and least biased estimation of the crystal structure.

4.1 Information from Diffraction Data

Conventional crystallographic data are usually in the form of structure amplitudes with phases which are unknown or estimated by techniques such as isomorphous replacement, anomalous dispersion, dynamical scattering effects, traditional direct methods, *etc*. Following the approach outlined by Gull and Daniell, (15) (see also Ref. 1) we introduce the information from diffraction measurements into p^{MI} via weak constraints, since this will help to guarantee a solution to the MI determination of structure and will also lead to some filtering of the structure consistent with the accuracy of the data. The constraints which we consider are:

4.1.1 Phases Approximately Known

Given N_1 distinct measured unitary structure factors, E_k, whose phases are at least partially known and for which the corresponding total estimated standard deviations are $\sigma_{k,l}$, then an appropriate weak constraint for incorporating this information into p^{MI2} is

$$f_1(p) \equiv \frac{1}{2N_1} \sum_{k \in D_1} |U_k - F_k|^2 / \sigma_{k,l}^2 = C_1, \tag{10}$$

Information & Crystal-Structure Estimation

where the summation is over the set D_1 of all measured reflexions including Friedel pairs and U_k is the model unitary structure factor corresponding to p and is given by (2). Constraint (10) represents the reduced χ^2 distribution when the errors in E_k are normally distributed, and should be s.t. $C_1 \simeq 1$ when the structure fits the data within experimental errors.

4.1.2 Phases Unknown

In this case, only the $|E_k|$ are assumed known, and an appropriate constraint is

$$f_2(p) \equiv \frac{1}{2N_2} \sum_{k \in D_2} |U_k - |E_k|\exp\{i\phi_k\}|^2 / \sigma_{k,2}^2 = C_2, \qquad (11)$$

where there are N_2 distinct reflexions with unknown phases belonging to the set D_2 and ϕ_k is the phase of U_k. This term does not have a χ^2 distribution for f_2 with $C_2 \simeq 1$.

An alternative constraint to (11) for including phaseless data which has some advantages over (11) (especially in treating superimposed reflexions, and isomorphous-replacement data) is to work directly with the intensities $I_k = |E_k|^2$, *viz.*

$$f_3(p) = \frac{1}{2N_3} \sum_{k \in D_3} (U_k^* U_k - I_k)^2 / \sigma_{k,3}^2 = C_3. \qquad (12)$$

This form of constraint for introducing the information contained in phaseless data into p^{MI2} also has better mathematical properties than $f_2(p)$. (see Table 1 of Ref. 1).

4.2 Prior Information

In addition to the information obtained directly from diffraction measurements, one may also have some idea of the general shape of the molecules, say, from electron microscopy or Patterson-function techniques. Such information typically takes the form of an envelope function and may be incorporated into the MI estimate for p *via* the prior density q in (5). The function q in (5) has the effect that density p^{MI2} is most easily modified in regions where q is large and very difficult (*i.e.* much additional information is required) to modify in regions where q is small.

Incorporation of prior knowledge about the structure by this means does not, of itself, contain any proper expression of one's degree of belief in the relative elements of that prior structure, especially in regions where q is of intermediate magnitude. It therefore appears valuable to note here that a simple and tractable means for incorporating a measure of one's degree of belief in q is *via* the constraint

$$f_4(p) \equiv \frac{1}{N} \sum_{j=1}^{N} \frac{(p_j - q_j)^2}{\sigma_{j,4}^2} = C \qquad (13)$$

where the expected value for C_4 is $\simeq 1$. This constraint (or restraint) acts as a bounding function on p (see Fig. 1).

We believe that the constraint (13) offers a potentially very powerful means for incorporating such information as: (i) presence of regions of almost flat electron density in the unit cell due to the presence of poorly ordered solvent molecules, (ii) boundedness of p in regions surrounding structural information from an incomplete atomic model of the structure and (iii) improved local knowledge of the structure as the determination progresses. More generally, by the judicious use of (13) one may hope to overcome many of the obvious limitations of the simple model outlined earlier in Sec. 3. The fact that (13) is convex in p and leads to a purely local contribution to the MI equation for p_j (see eqn (14) below) means that it is both an easy constraint to incorporate into the method and is of the type considered in Ref. 1 to be particularly helpful in structure estimation.

4.3 The Fundamental Equations

Given information of the type outlined above, the MI method of inference involves minimizing (5) subject to the particular constraints (10), (11), (12) and (13), which problem may be treated by the well-known method of Lagrange multipliers and leads to the fundamental equations

$$p_j = q_j \exp\{-\lambda_o - \lambda \cdot f^1_{r,j}(p)\}, \tag{14}$$

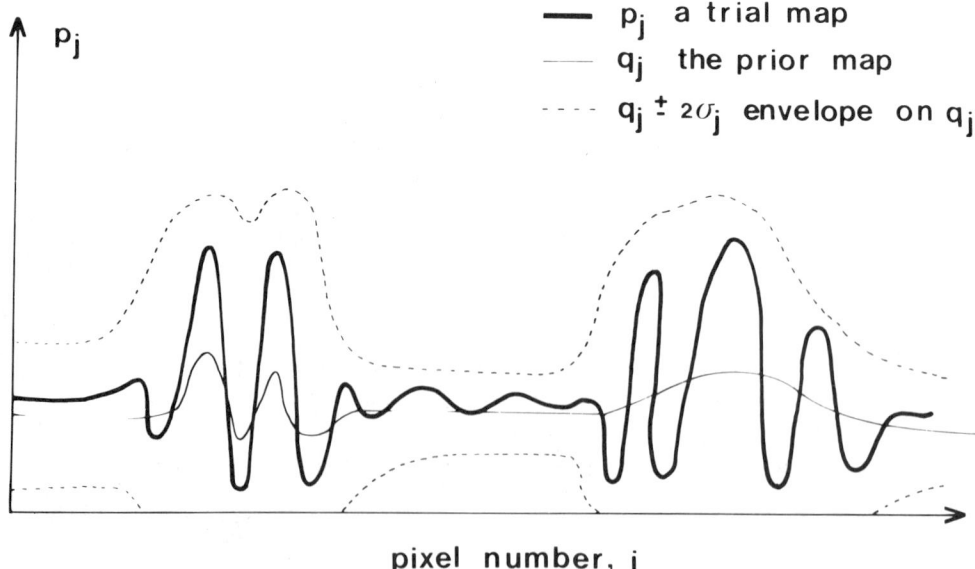

Figure 1. Schematic illustration of restraint (14) on the allowed deviation of a map, p, from a prior map, q. The surfaces $q \pm 2\sigma$ indicate the restraining surfaces about q.

where

$$f^1_{r,j}(p) = \frac{\partial f_r(p)}{\partial p_j},$$

λ_r is the associated Lagrange multiplier and λ_o is the Lagrange multipler arising from the normalization constraint on p. Solution of equations (14) for p gives stationary points of I, and so in principle one must always confirm that a solution gives a minimum of I, although in practice this may be obvious from the character of the solutions. The problem of numerical solution of (14) for different constraints is a large and difficult subject in itself and has been discussed by various authors, *e.g.* Refs. 17, 22, 23 and 24, and will not be considered here.

Conclusions

We have outlined some key elements of a general approach to X-ray structure estimation based on information theory. The proper implementation of these ideas offers to provide a very powerful means for the determination of crystal structures from X-ray diffraction data and other available information.

Acknowledgements

We are indebted to various colleagues for lively and constructive comments, and particularly to Drs. P. Colman, M.S. Lehmann, A.McL Mathieson, J. Skilling, and D. Stuart. This work was greatly aided by support from the Danish Science Foundation (SS) and a Visiting Fellowship from the Royal Society (SWW).

References and Footnotes

1. S.W. Wilkins, J.N. Varghese and M.S. Lehmann *Acta Cryst A39,* 47 (1983).
2. S.W. Wilkins, J.N. Varghese and S. Steenstrup in *Indirect Imaging,* Ed., J.A. Roberts, Cambridge Univ. Press, Cambridge p. 309 (1984).
3. S. Steenstrup and S.W. Wilkins, *Acta Cryst A40,* 163 (1984).
4. C.E. Shannon. *The Mathematical Theory of Communication, Bell Syst. Tech. J. 27* 399 (1948).
5. S. Kullback, *Annals of Math Statistics 22,* 79 (1951).
6. S. Kullback, *Information Theory and Statistics.* Wiley: New York (1959).
7. E.T. Jaynes, *Phys. Rev. 106,* 620 (1957).
8. I.J. Good, *Annals. Math. Stat. 34,* 911 (1963).
9. A. Wehrl, *Rev. Mod. Phys. 50,* 221 (1978).
10. A. Hobson, *Concepts in Statistical Mechanics.* Gordon and Breach, New York (1971).
11. A. Hobson and B-K Cheng, *J. of Stat. Phys. 7,* 301 (1973).
12. R.W. Johnson, *I.E.E.E. Trans. Inform. Theory IT-25* 129 (1979).
13. E.T. Jaynes, *I.E.E.E. Trans. Syst. Sci., Cybern. SSC-4* 227 (1968).
14. J.W. Shore and R.W. Johnson, *I.E.E.E. Trans on Inform. Theory IT-26,* 26 (1980).
15. S.F. Gull and G.J. Daniell, *Nature (Lond.) 272,* 686 (1978).
16. R.D. Levine, *J. Phys. A: Math. Nucl. Gen. 13,* 91 (1980).
17. G. Bricogne, *Acta Cryst. A40,* 410 (1984).
18. O. Piro, *Acta Cryst. A39,* 61 (1983).

19. H. Hauptmann, in *Direct Methods* ed. M.F.C. Ladd and R.A. Palmer. Plenum: New York (1980).
20. A. McLachlan, in *Proceedings of the CECAM-EMBO Workshop on Maximum Entropy and the X-Ray Phase Problem* (1984).
21. D. Sayre, in *Direct Methods.* ed. M.F.C. Ladd and R.A. Palmer. Plenum : New York (1980).
22. J. Skilling and R. Bryan, *Mon. Not. R. Astron. Soc.,* in the press.
23. S.W. Wilkins, *Acta Cryst. A39,* 892 and 896 (1983).
24. J. Navaza, *Acta Cryst. A. 41,* 232-244 (1985).

A Method for the Systematic Modification of Least-Squares Weights to Account for Residual Error

Wang Hong and Beverly E. Robertson
Faculty of Science, University of Regina
Regina, Saskatchewan, Canada S4S 0A2

Abstract

A method is discussed for the modification of least-squares weights by adding a quantity σ_m^2 to the calculated variance σ_c^2, of Δ ($|F_o| - |F_c|$). The quantity σ_m^2 is derived from the difference between the average value of σ_c^2 in a range of $|F_o|$ and $\sin\theta/\lambda$ and the value expected if σ_c^2 were the correct average variance of Δ. An estimation of the correct variance is obtained from a consideration of the distribution of the average value of $\omega_F \Delta^2$ where ω_F is the weight assigned to Δ. The procedure has been applied to the refinement of several crystal structures and the results for one structure are described in detail. The use of modified weights in least-squares refinement can lead to decreases in the calculated estimated standard deviations of the refined coordinates, from the estimated standard deviations of the coordinates resulting from refinements based only on σ_c^2, of up to 40%. A measure of chemical consistency has been used in an attempt to assess the real improvement in the refined coordinates. The results suggest that the real improvement in the refined coordinates is somewhat less than indicated by the apparent decrease in the estimated standard deviations of the refined coordinates.

Introduction

Nearly all crystal-structure determinations with data obtained from the diffraction of a single crystal are refined by least squares. The quantities minimized are either $\Sigma \omega_I (I_o - I_c)^2$ or $\Sigma \omega_F (|F_o| - |F_c|)^2$, where ω_I and ω_F are the weights to be assigned to the intensities I or the structure amplitudes $|F|$, respectively. A casual survey of recently published crystal-structure results indicates that the latter quantity, based on F, is still used in the majority of structure determinations (even to the extent that it is not always felt necessary to explicitly state which quantity is minimized).

The use of the least-squares method is formally based on the assumption that the n terms $\omega_F \Delta^2$ are independent, and the Δ have a joint distribution with zero means. The correct value of ω_F is the reciprocal of the variance of Δ, σ_F^2, and σ_F is the standard deviation of Δ. Although the use of the least-squares method does not have a theoretical rationale if these conditions are not met, they are seldom met fully in the data used for the refinement of the parameters which define a crystal structure. Furthermore, the true variances σ_F^2 are seldom known. Nevertheless, the history of crystallography suggsts that the method gives reasonably consistent results. We will take a pragmatic approach and attempt to obtain all the information contained in the experiment, even when the presence of various systematic errors changes the means of classes of Δ from zero.

Errors in Δ will be divided into those which are random and those which are systematic. In principle, systematic errors can be eliminated by accounting for them in the model used to calculate I_c (and $|F_c|$). This implies that the model allows a description of not only such sample-dependent effects as absorption, extinction and crystal degradation, but also accounts fully for the conditions of the experiment, such as slow changes in detector sensitivity, or room temperature, and inhomogeniety in the incident X-ray beam. In practice their presence may not be recognized, and even if it is recognized, the calculations to modify I_c may not be practicable. A useful list of important possible sources of random and systematic errors in I_o and $|F_o|$ has been given by Abrahams (2). Such lists should never be considered complete. Creative speculation about possible sources of significant error is not limited. For instance, errors resulting from systematic variation in scan speeds has received little attention. The interaction of a large irregular-shaped crystal with the typical inhomogeniety in a monochromatized X-ray beam has been observed to cause variations in the intensity of a particular reflection as diffraction geometry is varied of up to 15% (3).

Another class of errors associated with Δ are those arising from inadequacies of the physical model of the scattering electron density, which are by definition systematic (4). The latter have been discussed by Coppens (5). These errors are caused by differences between the electron density of the bonded atom and that of the free atom, including bond electron density. These differences can easily cause relative errors of 10% in the scattering factors of first row elements. The most well documented consequence of this systematic error is the substantially shorter length of bonds to hydrogen atoms as determined by X-ray diffraction, as compared to hydrogen bond lengths determined by neutron diffraction. The positions of first row elements may be similarly displaced, either toward lone-pair electron density or toward bonding electron density, by up to 0.015A (5).

In the sense that X-ray diffraction results are intended to describe the center of electron density, these shifts are not errors. If the coordinates of the centers of the electron density of the core electrons is the desired product of the structure determination then part of these shifts are indeed errors (6).

Wilson (7) has drawn attention to a systematic error in $|F_o|$ generated from the random error in I_o. The non-linear nature of the process of taking the square root of I_o to obtain $|F_o|$ means that the effects of positive errors in I_o on $|F_o|$ do not compensate the effects of negative errors in I_o. Wilson (8) has added that an approximate correction can be made by adding $1/8\, \sigma_I^2\, I_o^{-3/2}$ to $|F_o|$ if σ_I is known. The error in $|F_o|$ is greatest for weak intensities but the correction is less valid for weak intensities, suggesting a rationale for not using reflections for which $I_o < 3\sigma_I$ in refinements based on $|F_o|$.

Of the various sources of error which have been discussed or inferred above, the one which is most easily approximated is that which results form the Poisson distribution in the arrival of quanta in the incident X-ray beam. The true distribution for the number of counts representing I_o is the difference between the Poisson distribution for the total count and the Poisson distribution for the background count. [The difference does not have a Poisson distribution (9).] The true distribution is normally approximated by a Gaussian distribution. In the majority of reports of the determination of crystal structures, the contribution of counting statistics, in its approximate form, is the only error in Δ acknowledged in the calculation of weights for least squares.

Abrahams (10) has shown that the fractional contribution of counting statistics to the total variance predominates only for weak reflections, even when great care is taken with the experimental conditions (11).

Many attempts to modify weighting schemes have involved improved estimates of σ_I^2 from a careful analysis of the experimental conditions. Others involve the measurement of several members of a form for each reflection (10). This is not practical for data collected from crystals with very low symemtry or on diffractometers which meet heavy demands for data. Nevertheless such a procedure could probably be used more often than it is at present.

The ratio of $\Sigma \omega_F \Delta^2$ to $n-m$ (n was previously defined as the number of observations and m is the number of least-squares variables) is an indication of the validity of the weighting scheme. The square root of this ratio, S, is the goodness-of-fit and S should approach unity, with relatively high probability in the case of single-crystal structure refinements (12, 13). In order to bring S closer to unity it is common practice to add terms proportional to $|F_o|^2$ or $\sin \theta/\lambda$ to σ_F^2. The rationale is often given in terms of attempts to account for additional random or systematic error. The parameters which are the constants of proportionality are chosen either on an *ad hoc* basis or on a *trial-and-error* basis and tested by calculating $\langle \Sigma \omega_F \Delta^2 \rangle$ for ranges of $|F_o|$ and $\sin \theta/\lambda$ (14,15).

Weighting schemes based on parametric expansions are those of Cruickshank (16) and Nielsen (17). The latter minimized $\Sigma \omega_F \Delta^2 \log(\omega_F \Delta^2)$, with the condition that $\Sigma \omega_F \Delta^2$ is constant, by adjusting the coefficients in the expression for ω_F;

$$\omega_F = (a + b|F_o| + c|F_o|^2 + d \sin \theta/\lambda)^{-1}. \tag{1}$$

The consequences are that $\langle \Sigma \omega_F \Delta^2 \rangle$ is made reasonably independent of $|F_o|$ and $\sin \theta/\lambda$, and S approaches unity.

Such procedures either implicitly assume that systematic errors are absent, or that they can be treated in the same manner as random errors. An objection to these procedures is that if the parameters of the weighting scheme are adjusted by some minimization process, we have added to the number of variables that must be determined by the original data.

A further cause for concern has been discussed by Hamilton (18) which we restate as follows:

If a systematic error in $|F_o|$ corresponds to a systematic error in $|F_c|$ associated with one of the structural variables p_i, the error in $|F_o|$ could be partly obscured by adjusting p_i, and thus adjusting $|F_c|$ toward $|F_o|$ to minimize Δ^2 for each member of the group of reflections with systematic error. (This is the usual potential consequence of systematic error.) However if the Δ^2 have already been made artificially small for that group of reflections by creating errors in p_i, the adjustment of the weighting parameters may assign larger new relative weights than the original assignment based on the experimental conditions. Subsequent least-squares refinement with the new weights would then enhance the original error in p_i.

In order to determine if this actually happens, we will carry out trial refinements with data sets showing obvious systematic error. We will modify the variance calculated from a knowledge of the experimental conditions, which we will designate as σ_c^2, by adding a component σ_m^2, such that $\sigma_F^2 = \sigma_c^2 + \sigma_m^2$. Individual reflections with similar values of $|F_o|$ and $\sin\theta/\lambda$, or other parameters, may nevertheless have quite different values of σ_c because, for instance, of variations in background, peak shape, or counting time. We wish to avoid the loss of that information.

We have taken the generally accepted view that systematic error is any error that could be accounted for, at least in principle, by mathematical modifications to the model including the parameters that describe the experimental conditions. If such errors are subject to mathematical description it becomes possible to make predictions about the value of one Δ, given knowledge of another, and the Δ are therefore correlated. An easily recognized correlation is one in which all Δ is a given range of $|F_o|$ or $\sin\theta/\lambda$ are of the same sign; and therefore the Δ in that range do not have zero means. However, if the mathematical relations which might be used to account for systematic error are complex, the effect of such systematic errors on Δ, or a significant part of such effects, may be indistinguishable in a practical sense from random error.

Flack and Vincent (19) have shown that in the presence of correlation of a purely serial nature (systematic error associated with the data collection sequence), least squares still gives an unbiased estimate of the structural parameters, although the variance of the parameters may be in error.

Abrahams (2) has investigated the effects of various types of artificially generated systematic error in structural results. He used weights which were the reciprocal of the real variances arising from the random component of error in Δ. The results showed that for all types of systematic error which were generated (which were systematic error proportional to (i) I_o, (ii) the Bragg angle θ, (iii) the data-collection sequence number, and (iv) systematic error in absorption corrections) significant changes occurred in the derived scale factors and temperature factors. The values of R, ωR and S were significantly affected only in the case of systematic error of a serial nature (type iii) and only careful consideration of S could be used to infer that the error was systematic.

The surprising result was that the maximum deviations in the positional parameters from their correct values did not differ from those expected for the case of only random error in Δ in any of the trial refinements. Most structure determinations are carried out primarily to determine the positional parameters.

We conclude from the foregoing that it might not be particularly dangerous to include the contribution of systematic error to Δ in calculating the weights ω_F and we will set $\omega_F = \sigma_F^{-2}$, where σ_F represents all of the error in Δ, random and systematic, and examine the consequences.

For most structure determinations, the quantity σ_m^2, which will be added to σ_c^2 to form σ_F^2, can be expected to include substantial contributions from errors that are purely random in origin. It is possible to detect additional random errors beyond those associated with the counting statistics of individual reflections by observing the extent by which fluctuations in the standard reflections used to monitor the data collection exceed fluctuations expected from individual counting statistics. The same excess fluctuations may be assumed to affect the variance of the individual reflections. In practice the contribution can easily be as large as those arising from individual counting statistics. Such corrections are possible with some crystallographic program systems such as XRAY 76 (20). However they appear to be seldom applied, in which case they should appear in σ_m^2.

Calculation of the Modified Variance

It is well known that if the σ_F are correctly determined, and represent random error, $\Sigma \Delta^2/\sigma_F^2$ should approach $n-m$. Then the average value of the n quantities Δ^2/σ_F^2 would be $(n-m)/n$. Although the probability that this will be true for subsets of the data is lower, $\langle \Delta^2/\sigma_F^2 \rangle$ should not show systematic trends (bias) with respect to any parameters. It usually does so with respect to $|F_o|$ and $\sin \theta/\lambda$. We wish to choose σ_m^2 so that

$$\langle \Delta^2/\sigma_F^2 \rangle = (n-m)/n \qquad (3)$$

where the average is over reflections with common F_o and $\sin \theta/\lambda$.

We wish to replace the average of the ratio $\langle \Delta^2/\sigma_F^2 \rangle$, with the ratio of the average; ie, with $\langle \Delta^2 \rangle/\langle \sigma_F^2 \rangle$. However these quantities are not equal, because negative deviations from the average ratio overcompensate positive deviations. The approximate relationship between the average of the ratios and the ratio of the averages is given in textbooks on statistics as:

$$\langle \frac{\Delta^2}{\sigma_F^2} \rangle = \frac{\langle \Delta^2 \rangle}{\langle \sigma_F^2 \rangle} - \frac{1}{\langle \sigma_F^2 \rangle^2} \text{cov}[\Delta^2, \sigma_F^2] + \frac{\langle \Delta^2 \rangle}{\langle \sigma_F^2 \rangle^3} \text{var}[\sigma_F^2]. \qquad (4)$$

Combining eq. 2, 3, and 4,

$$\langle \sigma_m^2 \rangle = \frac{n}{n-m} \{\langle \Delta^2 \rangle - \frac{1}{\langle \sigma_F^2 \rangle} \text{cov}[\Delta^2, \sigma_F^2] + \frac{\langle \Delta^2 \rangle}{\langle \sigma_F^2 \rangle^2} + \text{var}[\sigma_F^2]\} - \langle \sigma_c^2 \rangle, \qquad (5)$$

In order to calculate $\langle \sigma_m^2 \rangle$ the data are divided into groups with common values of $|F_o|$ and $\sin \theta / \lambda$ and the right-hand side of eq. 5 is estimated. The quantities σ_F^2 are not known but the best available estimate of σ_F^2 is its unmodified value, σ_c^2. The points represented by eq. 5 can then be fitted by a general series expansion in terms of the variables V_θ and V_F where

$$V_\theta = \frac{\sin \theta}{\sin \theta_{max}}, \qquad V_F = \frac{|F_o|}{|F_o|_{max}}. \qquad (6)$$

The series expansion, up to order p, is then

$$\langle \sigma_m^2 \rangle = \sum_{q=0}^{p} \sum_{l=0}^{q} A_{q-l,l} V_\theta^{q-l} V_F^l \qquad (7)$$

where $A_{q-l,l}$ are the series coefficients.

Eq. 7 is used to calculate σ_m^2 and the modified weights after the $A_{q-l,l}$ have been determined by a least-squares fit to $\langle \sigma_m^2 \rangle$. The structure is then refined to completion, also by least squares, using the modified weights. The Δ will be changed by the least squares and it will be necessary to iterate the procedure until it converges; i.e., until σ_m^2 and Δ are nearly constant. In successive iterations improved values of σ_F^2 will be available for use in the right-hand side of eq. 5.

Results

In order to test the proposed method of modifying weights, further approximations have been made and the method has been tested on a number of recently determined structures. The approximations were:

   (i) The maximum power p was limited to 2, which generates six $A_{q-l,l}$ values,
   (ii) The variance and covariance terms in eq. 5 have been ommitted,
   (iii) The quantity m has been set to zero in eq. 5,
   (iv) In the process of averaging $\langle \sigma_c^2 \rangle$ we have taken the root-mean-square average rather than the true mean.

The implications of the second and third approximation are that σ_m^2 are underestimated. The implications of the fourth approximation are that reflections that deviate far from the mean are overweighted in their contribution to the average. The advantage of these approximations is that the calculation now reduces to fitting the series expansion to $\Delta^2 - \sigma_c^2$ plotted directly as a function of $|F_o|$ and sin θ/λ, without averaging $\Delta^2 - \sigma_c^2$.

We wish to determine if: (i) the method enhances the effects of systematic error; (ii) it converges; (iii) it improves the accuracy of structural parameters. We would not propose to answer the last question only by comparing the estimated standard deviations (e.s.d.'s) of the least-squares variables at the end of refinement. Even if we can justify the method of weight modification in terms of pragmatism, we have not attempted to justify the validity of the derived e.s.d.'s. Furthermore, the quantities of ultimate interest are themselves functions of the least-squares variables and their accuracy depends not only on the e.s.d.'s of the least-squares variables, but also on the covariance of the least-squares variables.

In order to make an independent assessment of the relative accuracy of two refinements of the same structure using different weights, a consistency index C was calculated. If q_j ($j=1, 2, ...r$) is the number of chemically equivalent (or chemically similar) bonds of type j, for all q_j bonds of type j we assign C_j where

$$C_j^2 = \sum_{i=1}^{q_j} (b_{ij} - \langle b_j \rangle)^2 / (q_j - 1) \tag{8}$$

and b_{ij} is the length of the ith bond of type j and $\langle b_j \rangle$ is the average length of all bonds of type j. C is the average of the $\sum_{j=1}^{r} q_j$ values of C_j.

All data were collected on a modified Picker FACS-I diffractometer with MoKα radiation, a highly oriented graphite monochromator and the NRCC data-collection program package operating in the peak-profile mode (21). Background was counted at each limit of the scan for a minimum of 20% of the peak-scan time. Major computing, including refinement, was carried out on a Honeywell Sigma 9 computer using the XRAY 76 system of programs (20). The variance calculated for the intensity of each reflection, σ_I, was determined from the counting statistics of that reflection with an additional contribution derived from the excess scatter in the standard reflections. The calculated variance in F was estimated from (22).

$$\sigma_c^2 = (\sqrt{I + \sigma_I} - \sqrt{I})^2. \tag{9}$$

Both refinements of a structure used the same data which were that with $I > n_I \sigma_I$ (23).

Table I shows the crystal data, intensity measurement and refinement data for cocaine (24). Table II compares the values of R, ωR, S and C, and lists the percentage fractional average decrease in the estimated standard deviations of the least-squares variables. Table II also shows the average ratio of the difference between least-squares variables from the two refinements to the estimated standard deviation obtained with modified weights. Table III shows the distribution of $\langle \omega_F \Delta^2 \rangle$ and R with respect to $|F_o|$ and $\sin \theta/\lambda$. The intensity data were not corrected for the effects of absorption or secondary extinction. The original refinement with a σ_c^{-2} weighting scheme showed slight positive bias with respect to $|F_o|$ and strong negative bias with respect to $\sin \theta/\lambda$; ie, $\langle \omega_F \Delta^2 \rangle$ increased slightly with increasing $|F_o|$ and decreased strongly with increasing $\sin \theta/\lambda$. The process of weight modification was iterated three times. There was no clear evidence of systematic error in the data.

In some cases σ_m^2 was calculated to be a negative quantity suggesting that the average value of σ_c^2 in that range of $|F_o|$ and $\sin \theta/\lambda$ was originally overestimated. In these cases σ_m^2 was reset to zero.

The goodness-of-fit has been reduced by the application of weight modification from 2.92 to 1.13. Because of the approximations made, the minimum value predicted is not 1.00 but 1.09. R has not been changed but ωR is reduced from 0.062 to 0.050. The e.s.d.'s in the refined variables have been reduced by a total of approximately 40%, which could, in theory, be achieved by doubling the time spent collecting data.

The average percentage decrease in e.s.d.'s of the refined parameters was 33% on the first application of weight modification and 6% and 1% on two successive iterations. Small changes occurred in the coefficients $A_{q-l,l}$ on successive iterations. The average ratio of the magnitude of changes in the parameters to their e.s.d's show that the structure determined with modified weights is significantly different from the original. Variation in the percentage improvement in e.s.d.'s suggest that the residual errors represented by σ_m^2, whether they are random or systematic, are nevertheless anisotropic.

The molecule of cocaine has low internal symmetry and the consistency index is averaged over only 14 bond lengths, which include a phenyl ring and four pairs of similar bonds which are not chemically equivalent with respect to a side group of the molecule. Nevertheless, the small decrease in C is less than that which might have been expected from the large improvement in the e.s.d.'s. The results shown in Table III indicate that the method has succeeded in removing bias from the weights.

Conclusion

The results for cocaine and for other trial refinements clearly indicate that the method converges. In each case it was found that the e.s.d's were substantially reduced although not necessarily by as much as in the case of cocaine. The reduction in the consistency index indicates, especially for the case of molecules showing high internal symmetry and clearly evident systematic error, that the improvement in accuracy is not as great as that indicated by the improvement in the e.s.d.'s. We suggest that if any systematic error is included in σ_m^2, the value of SM_{ii}^{-1} (where M

Least-Squares Weights & Residual Errors

Table I
Data collection, crystal data and refinement parameters for cocaine

Formula	$C_{17}H_{21}NO_4$	$\mu(MoK\alpha)$	0.0958 cm^{-1}
Crystal dimensions	0.7 × 1.0 × 0.2 mm	a	10.130(1)Å
$\sin\theta/\lambda$	0.7048Å^{-1}	b	9.866(2)Å
$|F_o|_{max}$	55.9	c	8.445(1)Å
n	1818	β	106.92(1)°
m	283	A_{00}	3.050
n_I	1.3	A_{10}	−10.222
Original weights	σ_c^{-2}	A_{20}	7.881
Space Group	$P2_1$	A_{01}	−10.584
Mol. wt	303.4	A_{02}	11.111
Z	2	A_{11}	20.965

Table II
Comparison of original refinement and refinement with modified weights

	Original	Modified Weights
R	0.065	0.065
ωR	0.062	0.050
S	2.90	1.13
C	0.0100Å	.0092Å
	$\dfrac{100\langle\sigma_{orig} - \sigma_{mw}\rangle}{\sigma_{orig}}$	$\dfrac{\langle|P_{orig} - P_{mw}|\rangle}{\sigma_{mw}}$
x	36	1.64
y	40	1.20
z	34	1.95
U_{11}	39	1.80
U_{22}	46	1.19
U_{33}	35	1.71
U_{12}	44	1.40
U_{13}	37	1.65
U_{23}	41	1.00

is the matrix of the normal equations) must be treated as an estimate of the minimum value of the e.s.d. of the i'th least-squares variable.

Other trial refinements included examples which showed clear systematic error, such as consistently negative values of Δ for large $|F_o|$, etc. In each case, the reflections in question were weighted lower relative to other reflections as a result of the modifications of the weights; in other words, σ_m^2 for these reflections was large. We conclude that it is unlikely that the method will amplify the effects of systematic error.

Provided that the numbers now reported as e.s.d's are interpreted instead as minimum quantities, the method appears to be superior to present methods of modify-

Table III
Distribution of $\langle \omega_F \Delta^2 \rangle$

(a) with respect to $|F_o|$

$|F_o|$ interval	$\langle \omega_F \Delta^2 \rangle$	$100R$
0-1.58	0.58	20
1.58-1.74	1.26	23
1.74-1.89	1.10	25
1.89-2.04	0.78	21
2.04-2.19	1.11	25
2.19-2.35	0.96	21
2.35-2.56	0.75	18
2.56-2.76	1.70	13
2.76-3.00	1.25	12
3.00-3.35	1.46	16
3.35-3.69	1.24	12
3.69-4.13	0.99	6
4.13-4.68	1.25	5
4.68-5.35	1.20	3
5.35-6.22	1.31	3
6.22-7.79	0.91	3
7.79-9.20	0.74	3
9.20-12.51	0.56	2
12.51-19.69	1.08	3
19.69-∞	1.18	3

(b) with respect to $\sin\theta/\lambda$

$\sin\theta/\lambda$	$\langle \omega_F \Delta^2 \rangle$	$100R$
0-0.230	0.99	3
0.230-0.290	1.05	3
0.290-0.333	1.26	3
0.333-0.369	1.00	2
0.369-0.400	1.13	2
0.400-0.427	1.13	3
0.427-0.449	1.03	3
0.449-0.473	1.36	4
0.473-0.494	1.28	5
0.494-0.513	0.89	5
0.513-0.535	1.29	7
0.535-0.550	1.87	9
0.550-0.571	0.93	10
0.571-0.591	0.75	12
0.591-0.609	0.43	11
0.609-0.628	0.85	19
0.628-0.648	0.86	23
0.648-0.668	0.79	23
0.668-0.680	1.26	33
0.680-∞	1.32	42

Each interval of $|F_o|$ and $\sin\theta/\lambda$ contains approximately 90 reflections.

ing weights by adding so-called *ignorance factors* such as $k|F_o|^2$ to σ_c^2 on an *ad hoc* basis.

The experience to date suggests several improvements are possible. In particular for some data sets cubic terms are necessary in the expansion for $\langle \sigma_m^2 \rangle$. The variance and covariance terms in eq. 5 should be included in the calculations and the factor $(n-m)/n$ should not be set equal to unity. Finally, the true average of the right-hand side rather than the root-mean-square value should be used. A program will be available as part of the XTAL system which incorporates these changes.

The method has certain characteristics which are similar to those of the robust/resistant method (25). However, there are significant differences. Weight modification does not increase the radius of convergence of the refinement. The robust/resistant method may do so. Weight modification changes the relative weights of groups of reflections, not individual reflection as does the robust/resistant method. If the residuals R and ωR are small, the robust/resistant method will give essentially the same result as general least squares based on σ_c^2. Modified weights will lead to significantly different results because the weights will be different even for low residuals. It will be necessary for the crystallographer to consider the problem at hand and choose between the robust/resistant algorithm and the weight modification procedure.

The authors thank the Canadian International Development Agency for the support of W. H., the University of Regina for the provision of computer services, A. G. Law, R. J. Barton, R. S. Coleman and A. Adatia, University of Regina, for helpful discussions, and the Natural Sciences and Engineering Research Council for an Operating Grant.

References & Footnotes

1. W. C. Hamilton, *Statistics in Physical Science,* Ronald Press, New York, p. 124 (1964).
2. S. C. Abrahams, *Acta Cryst. A25,* 165 (1969).
3. J. S. Rollet in *Computational Crystallography,* Ed., D. Sayre, Clarendon Press, Oxford, p. 338 (1982).
4. S. Harkema, J. Dam, C. J. van Hummel and A. J. Reuvers, *Acta Cryst. A 36,* 433 (1980)
5. P. Coppens, *Acta Cryst. A25,* 180 (1969)
6. B. E. Robertson and J. J. Stezowski, *Acta Cryst. B34,* 3005 (1978).
7. A. J. C. Wilson, *Acta Cryst. A32,* 994 (1976)
8. A. J. C. Wilson, *Acta Cryst. A35,* 122 (1979)
9. A. J. C. Wilson, *Acta Cryst. A36,* 929 (1980)
10. S. C. Abrahams, *Acta Cryst. B30,* 261 (1974)
11. Strictly speaking, the quantity with which Abrahams compared the variance derived from counting statistics was not a variance because it probably included systematic error.
12. It should be noted that S^2 or S can be made to approach unity by rescaling ω_F, by rescaling σ_F. However, R, ωR and the least-squares minimum will not be altered.
13. The estimated standard deviations, σ_p, in the variables p, on which F_c depends, are given by $\sigma_{pi} = M_{ii}$ where M is the matrix of the coefficients of the normal equations. This assumes $S \simeq 1$. Because σ_F and ω_F are not known, S usually deviates significantly from 1 and σ_{pi} is compensated by multiplying it by S.

14. K. Sudarsanan and R. A. Young, *Acta Cryst. B30*, 1381 (1974).
15. The common name for this modification is the *ignorance factor*.
16. D. W. J. Cruickshank in *X-ray Crystal Analysis*, Ed. R. Pepinksy, J. M. Robertson and J. C. Speakman, Pergamon Press, New York, p. 45, (1961).
17. K. Nielson, *Acta Cryst. A33*, 1009 (1977).
18. W. C. Hamilton, *International Tables for X-ray Crystallography, Vol. IV*, Ed. J. A. Ibers and W. C. Hamilton, Kynoch Press, Birmingham, England, p. 293 (1974).
19. H. D. Flack and M. G. Vincent, *Acta Cryst. A36*, 495 (1980).
20. J. M. Stewart, P. A. Machin, C. W. Dickinson, H. L. Ammon, H. Heck and H. D. Flack, *The XRAY 76 System, Tech. Rep. TR446*, Computer Science Center, University of Maryland, College Park, Maryland.
21. A. C. Larson and E. J. Gabe. *Computing in Crystallography*, Ed. H. Schenk, R. Olthof-Hazekamp, H. van Koningsveld and G. S. Bassi, Delft University Press, p. 81, (1978).
22. A preferable expression is $\sigma_c = \dfrac{\sqrt{I + \sigma_I} - \sqrt{I - \sigma_I}}{2}$. This and eq. 9 are both approximations. However they give more realistic values of σ_c when σ_I/I is relatively large than the more familiar approximation, $\sigma_c = \left(\dfrac{F}{I}\right)\sigma_I = \dfrac{\sigma_I}{2\sqrt{I}}$
23. It may be shown that this expression also excludes data from the refinement for which $|F_o| < n_F \sigma_c$ and $\dfrac{1}{n_F} = \sqrt{1 + \dfrac{1}{n_I}} - 1$. If σ_c is calculated as $\dfrac{\sigma_I}{2\sqrt{I}}$, $n_F = 2n_I$
24. R. J. Hrynchuk, R. J. Barton and B. E. Robertson, *Can. J. Chem. 61*, 481 (1983).
25. W. L. Nicholson, E. Prince, J. Buchanan and P. Tucker in *Crystallographic Statistics*, Ed. S. Ramaseshan, M. F. Richardson and A. J. C. Wilson, p. 229, Indian Academy of Sciences, Bangalore (1982).

*Structure & Statistics in Crystallography, ISBN 0-940030-10-1,
Ed., A. J. C. Wilson F. R. S., Adenine Press, ©Adenine Press 1985.*

The Variances and Covariances of Measured Intensities in Precise Lattice-Constant Determination by the Bond Method

Ewa Gałdecka
Institute for Low Temperature and Structure Research
Polish Academy of Sciences
Plac Katedralny 1, 50-950 Wrocław, Poland

Abstract

A mathematical model of variances and covariances of measured intensities has been derived in which statistical errors of the following factors besides counting (Poisson) statistics were taken into account: angle setting, angle reading, primary beam intensity, sample temperature and counting time. This model was verified by the measurements on a Bond system diffractometer [W. L. Bond, *Acta Cryst. 13*, 814-818 (1960); erratum: *Acta Cryst.* A*31*, 698 (1975)]. It can also be used for the other diffractometers with step-scanning and a fixed-time count.

Introduction

Precision of the measurements on the Bond diffractometer depends on several factors as discussed in my papers (1,2). In theoretical considerations it is usually assumed that the variances $\sigma^2(h_i)$ of measured intensities (recorded numbers of counts) h_i, $i=1,2,\ldots,n$ (where n is the number of measuring points) are equal to the expected values of $\epsilon\{h_i\}$ of these intensities

$$\sigma^2(h_i) = \epsilon\{h_i\} \qquad (1)$$

and that the covariances $\text{cov}(h_i,h_j)$ between the intensities h_i, h_j in different points ($i \neq j$) of the diffraction profile are equal to zero

$$\text{cov}(h_i,h_j)=0, \qquad i \neq j. \qquad (2)$$

These two assumptions—resulting from the Poisson distribution of the recorded counts—lead to a simple and effective conclusions concerning variances of the estimated line-profile parameters (1,3,4), which can be used in order to derive from them some parameters of physical interest with their variances.

In practice, however, the equalities (1) and (2) are not always satisfied and often

$$\sigma^2(h_i) > h_i \tag{3}$$

and

$$\mathrm{cov}(h_i, h_j) \neq 0. \tag{4}$$

This means, that—besides the counting statistics—there are other factors, being random variables, influencing measured intensities, that cannot be neglected in general case. Wilson (5) discussed the influence of statistical errors in the angular scale and in the scanning mechanism on the variances of the intensities. The aim of this paper is to find other factors and to derive the more complete model of variances and covariances of measured intensities.

Assumptions and Preliminary Considerations

When each of M factors influencing a measured intensity has a probability distribution $p_m(h_i)$, $m=1,2,\ldots,M$, with a variance $\sigma^2(h_i)_m$ the resulting probability distribution $p(h_i)$ can be treated as a convolution of these M partial distributions:

$$p(h_i) = p_1(h_i) * p_2(h_i) * \ldots * p_M(h_i) \tag{5}$$

According to the central-limit theorem, when M is sufficiently large, $p(h_i)$ tends to the normal distribution and when—in addition—distributions $p_m(h_i)$ are independent, the variance $\sigma^2(h_i)$ of $p(h_i)$ can be written as:

$$\sigma^2(h_i) = \sigma^2(h_i)_1 + \sigma^2(h_i)_2 + \ldots + \sigma^2(h_i)_M. \tag{6}$$

I assume that the above conditions are well satisfied and the equation (6) may be applied in further considerations.

In formulating the mathematical model of variances and covariances of measured intensities the prerequisites described in the following paragraphs have been taken into account.

The counting statistics

Recorded number of counts is a random variable with the Poisson distribution, with the variance $\sigma^2(h_i)_p$ given by equation

$$\sigma^2(h_i)_p = h_{i,0} \tag{7}$$

where

$$h_{i,0} = \epsilon\{h_i\}.$$

Variances of Intensities in the Bond Method

I assume that there are no covariances between two records of the counter.

The normalization of the profile

The intensity h in the Bond method is a function of the sample position ω:

$$h = h(\omega) \tag{8}$$

In particular, since the profile $h(\omega)$ is measured in separate points ω_i, the values h_i of the intensities can be expressed as

$$h_i = h(\omega_i), \quad i = -p,...,p \tag{8a}$$

where $2p+1 = n$ is the number of measuring points.

It is convenient to normalize the values ω and h by substituting:

$$2(\omega - \omega_0)/\omega_h = x \tag{9}$$

and

$$h(\omega)/H = h[x(\omega)]/H = y \tag{10}$$

where x and y are new variables and ω_0, ω_h and H—parameters of the profile. In an ideal case, when $h(\omega)$ is completely symmetrical, ω_0 means the peak position, H—the peak height and ω_h—the half-maximum width.

The variances and the covariances of the sample positions

In the case of automatic scanning—used in the Bond diffractometer—the first sample position ω_{-p} is set by hand and all the next ones—by a stepping motor with the scanning step $\Delta\omega$. The sample position is controlled (read) at the points ω_{-p} and ω_p. The angle-reading error $\delta\omega_R$ with the variance $\sigma^2(\omega)_R$ as well as the angle-setting error $\delta\Delta\omega$ with the variance $\sigma^2(\Delta\omega)$ are sources of a dispersion of estimated angle positions ω_i.

Actual (not known) positions $\omega_{i,0}$, $i=-p,...,p$, are given by

$$\begin{aligned}\omega_{i,0} &= \omega_{-p,0} + \sum_{k=1}^{i+p} \Delta\omega_k \\ &= \omega_{-p,0} + (i+p)\Delta\omega_N + \sum_{k=1}^{i+p} \delta\Delta\omega_k\end{aligned} \tag{11}$$

where $\Delta\omega_N$ is a nominal and $\Delta\omega_k$, $k=1,2,...,2p$ are actual values of scanning step and $\delta\Delta\omega_k = \Delta\omega_k - \Delta\omega_N$ are the angle-setting errors.

The estimated sample positions, ω_i, can be presented as:

$$\begin{aligned}
\omega_{-p} &= \omega_{-p,0} + \delta\omega_{-p,R} \\
\omega_i &= \omega_{-p} + (i+p)\Delta\hat{\omega}, \quad i = -p+1,\ldots,p-1 \\
\omega_p &= \omega_{p,0} + \delta\omega_{p,R}
\end{aligned} \quad (12)$$

where $\delta\omega_{-p,R}$ and $\delta\omega_{p,R}$ are the angle-reading errors of the first and last sample position respectively and

$$\Delta\hat{\omega} = (\omega_p - \omega_{-p})/(2p) \quad (12a)$$

is the mean value of the scanning step.

From (12a) and (12) results:

$$\begin{aligned}
\Delta\hat{\omega} &= (\omega_{p,0} + \delta\omega_{p,R} - \omega_{-p,0} - \delta\omega_{-p,R})/(2p) \\
&= (\omega_{-p,0} + 2p\Delta\omega_N + \delta\omega_{p,R} - \omega_{-p,0} - \delta\omega_{-p})/(2p) \\
&= \Delta\hat{\omega}_N + (\delta\omega_{p,R} - \delta\omega_{-p,R})/(2p) + 1/(2p)\sum_{k=1}^{2p}\delta\Delta\omega.
\end{aligned} \quad (13)$$

The error $\delta\omega_i$ of the estimated position ω_i is equal to the difference between ω_i and $\omega_{i,0}$:

$$\begin{aligned}
\delta\omega_i &= \omega_i - \omega_{i,0} \\
&= \delta\omega_{-p,R} + (i+p)(\Delta\hat{\omega} - \Delta\omega_N) - \sum_{k=1}^{i+p}\delta\Delta\omega_k \\
&= \delta\omega_{-p,R} + (i+p)/(2p)(\delta\omega_{p,R} - \delta\omega_{-p,R}) \\
&\quad + (i+p)/(2p)\sum_{k=1}^{2p}\delta\Delta\omega_k - \sum_{k=1}^{i+p}\delta\Delta\omega_k \\
&= (p-1)/(2p)\,\delta\omega_{-p,R} + (i+p)/(2p)\delta\omega_{p,R} \\
&\quad + (i+p)/(2p)\sum_{k=i+p+1}^{2p}\delta\Delta\omega_k + (i-p)/(2p)\sum_{k=1}^{2p}\delta\Delta\omega.
\end{aligned} \quad (14)$$

The variance $\sigma^2(\omega_i)$ of ω_i can be calculated from the definition:

$$\begin{aligned}
\sigma^2(\omega_i) &= \epsilon\{(\omega_i - \omega_{i,0})^2\} \\
&= \epsilon\{(\delta\omega_i)^2\}.
\end{aligned} \quad (15)$$

Variances of Intensities in the Bond Method 141

Assuming that there are no correlations between $\delta\omega_{p,R}$ and $\delta\omega_{-p,R}$, $\delta\omega_{p,R}$ and $\delta\Delta\omega_k$, $k=1,...,2p$, $\delta\omega_{-p,R}$ and $\delta\Delta\omega_k$, $k=1,...,2p$ as well as $\delta\Delta\omega_k$ and $\delta\Delta\omega_l$, $k\neq l$, $k,l=1,2,...,2p$, we obtain finally from (15) and (14):

$$\sigma^2(\omega_i)=(1+i^2/p^2)\sigma^2(\omega)_R/2+p(1-i^2/p^2)\sigma^2(\Delta\omega)/2. \tag{16}$$

As the result of angle-setting method there appear also the covariances between the positions ω_i and ω_j, $i\neq j$, $i,j=-p,...,p$. From the definition

$$\text{cov}(\omega_i,\omega_j)=\epsilon\{(\omega_i-\omega_{i,0})(\omega_j-\omega_{j,0})\}$$
$$=\epsilon\{\delta\omega_i\delta\omega_j\} \tag{17}$$

and from (14) results:

$$\text{cov}(\omega_i,\omega_j)=(1+ij/p^2)\sigma^2(\omega)_R/2+p(1-ij/p^2)\sigma^2(\Delta\omega)/2$$
$$-|i-j|\sigma^2(\Delta\omega)/2. \tag{18}$$

The primary beam intensity

The level of measured intensities h_i given by H [formula (10)] depends on the primary beam intensity I, which is a random variable with the mean value I_0 and the variance $\sigma^2(I)$ and depends on various factors (for example—on fluctuations of the power voltage).

I assume that the values H and I are proportional

$$H=c_I I \tag{19}$$

where c_I is a coefficient of proportionality. From (19) and (10) we obtain

$$h=c_I I y \tag{20}$$

The counting time

In the case of fixed-time counting, used in the Bond method, the intensities h_i are proportional to the counting rates r_i

$$h_i=\tau r_i \tag{21}$$

where τ is the counting time, which can also be treated as a random variable with the mean value τ_0 and the variance $\sigma^2(\tau)$.

The temperature of the sample

The intensities h_i depend on the temperature T—a random variable with the mean value T_0 and the variance $\sigma^2(T)$. For simplification I assume that the temperature influences only the peak position ω_0 and doesn't influence the peak height H and the peak shape $y(x)$, so that

$$h = h[\omega_0(T)] \tag{22}$$

and—additionally—that the dependence of ω_0 and T is linear:

$$\omega_0(T) = \Omega_0 + c_T T \tag{23}$$

where $\Omega_0 = \omega_0(T_0)$ and c_T is a coefficient of proportionality.

A Mathematical Model of the Variances and Covariances of Measured Intensities

Combining all the factors influencing measured intensities, described in the above sections, and taking into account the formula (6) we can express the variance $\sigma^2(h_i)$ as a sum of partial variances

$$\sigma^2(h_i) = \sigma^2(h_i)_p + \sigma^2(h_i)_\omega + \sigma^2(h_i)_I + \sigma^2(h_i)_\tau + \sigma^2(h_i)_T \tag{24}$$

in which $\sigma^2(h_i)_p$ is a component connected with the counting (Poisson) statistics given by (7) and each of the remaining components refers to the factor mentioned as an index, i.e. to ω, I, τ and T respectively.

To find the values of $\sigma^2(h_i)_\omega$, $\sigma^2(h_i)_I$, $\sigma^2(h_i)_\tau$ and $\sigma^2(h_i)_T$ we apply the principle of error propagation. In result the formula (24) can be written as

$$\sigma^2(h_i) = h_{i,0} + \left(\frac{\partial h}{\partial \omega}\right)_i^2 \sigma^2(\omega_i) + \left(\frac{\partial h}{\partial I}\right)_i^2 \sigma^2(I)$$
$$+ \left(\frac{\partial h}{\partial \tau}\right)_i^2 \sigma^2(\tau) + \left(\frac{\partial h}{\partial T}\right)_i^2 \sigma^2(T) \tag{25}$$

where $\left(\frac{\partial h}{\partial \omega}\right)_i$, $\left(\frac{\partial h}{\partial I}\right)_i$, $\left(\frac{\partial h}{\partial \tau}\right)_i$ and $\left(\frac{\partial h}{\partial T}\right)_i$ are partial derivatives in the point $h_{i,0} = h(\omega_{i,0}, I_0, \tau_0, T_0)$.

Using the dependences given above we calculate the partial derivatives. For simplification we denote:

$$\left(\frac{\partial h}{\partial \omega}\right)_i = h'(\omega_{i,0}) \tag{26}$$

From (20) we have

$$\left(\frac{\partial h}{\partial I}\right)_i = c_I y_{i,0} = h_{i,0}/I_0 \tag{27}$$

Variances of Intensities in the Bond Method

where

$$y_{i,0}=y[x(\omega_{i,0})].$$

From (21) results

$$\left(\frac{\partial h}{\partial \tau}\right)_i = r_{i,0} = h_{i,0}/\tau_0. \tag{28}$$

Combining (22) and (23) we obtain

$$\frac{\partial h}{\partial T} = \frac{h}{\omega_0}\frac{d\omega_0}{dT} = \frac{\partial h}{\partial \omega_0}c_T. \tag{29}$$

It is convenient to express $\frac{\partial h}{\partial \omega_0}$ with the aid of $\frac{\partial h}{\partial \omega}$, using (9) and (10):

$$\frac{\partial h}{\partial \omega_0} = \frac{\partial h}{\partial x}\frac{\partial x}{\partial \omega_0} = \frac{\partial h}{\partial \omega}\frac{\partial \omega}{\partial x}\frac{\partial x}{\partial \omega_0} = \frac{\partial h}{\partial \omega}\frac{1}{\frac{\partial x}{\partial \omega}}\frac{-2}{\omega_0}$$

$$= \frac{\partial h}{\partial \omega}\frac{\omega_h}{2}\frac{-2}{\omega_0} = -\frac{\partial h}{\partial \omega}. \tag{30}$$

Combining (29) with (30) and taking (26) into account we obtain

$$\frac{\partial h}{\partial T}_i = -\frac{\partial h}{\partial \omega}_i c_T = -h'(\omega_{i,0})c_T. \tag{31}$$

Substituting (26), (27), (28) and (31) into (25) and ordering the equation we obtain the final expression for variances:

$$\sigma^2(h_i) = h_{i,0} + h_{i,0}^2[\sigma^2(I)/I_0^2 + \sigma^2(\tau)/\tau_0^2] + h'^2(\omega_{i,0})[\sigma^2(\omega_i) + c_T^2\sigma^2(T)] \tag{32}$$

where $\sigma^2(\omega_i)$ is given by (16).

Similar considerations to that concerning the variances lead to the expression for covariances of the measured intensities. Taking into account the principle of error propagation and the assumptions described above we have:

$$\begin{aligned}\text{cov}(h_i,h_j) &= h_{i,0}h_{j,0}[\text{cov}(I_i,I_j)/I_0^2 + \text{cov}(\tau_i,\tau_j)/\tau_0^2] \\ {}_{i \neq j} \quad &+ h'(\omega_{i,0})h'(\omega_{j,0})[\text{cov}(\omega_i,\omega_j) + c_T^2\text{cov}(T_i,T_j)] \\ &= h_{i,0}h_{j,0}[\sigma^2(I)/I_0^2\rho(I_i,I_j) + \sigma^2(\tau)/\tau_0^2\rho(\tau_i,\tau_j)] \\ &+ h'(\omega_{i,0})h'(\omega_{j,0})[\text{cov}(\omega_i,\omega_j) + c_T^2\sigma^2(T)\rho(T_i,T_j)]\end{aligned} \tag{33}$$

where the ρ values are the correlation coefficients and $\text{cov}(\omega_i,\omega_j)$ is given by (18). According to the assumption given above the covariances between records of a counter have been neglected in (33).

The significance of the individual components in (32) and (33) cannot be discussed in detail without information concerning the diffractometer used and the stability of measuring parameters, *i.e.* without knowledge of variances and covariances of I, τ, T and $\sigma^2(\omega)_R$ and $\sigma^2(\Delta\omega)$ characterizing the class of the diffractometer and the quality of the measurement.

It seems however—taking into consideration the results obtained in my institute (6)—that the variances of angle-reading $\sigma^2(\omega)_R$ and angle-setting $\sigma^2(\Delta\omega)$ have—besides the counting statistics—the main influence on $\sigma^2(h_i)$ and $\text{cov}(h_i,h_j)$. In this case, when the variances and covariances of τ and T can be neglected, a simpler model can be proposed:

$$C_{i,j} = e_1\sqrt{h_{i,0}h_{j,0}} + e_2 h_{i,0} h_{j,0}$$
$$+ [e_3 + ij/p^2 e_4 - |i-j|e_5] h'(\omega_{i,0}) h'(\omega_{j,0}) \quad (34)$$

where

$$C_{i,j} = \sigma^2(h_i) \quad \text{for } i=j$$
$$= \text{cov}(h_i,h_j) \quad \text{for } i\neq j \quad (34a)$$

$$e_1 = 1 \quad \text{for } i=j$$
$$= 0 \quad \text{for } i\neq j \quad (34b)$$

$$e_2 = \sigma^2(I)/I_0^2 \text{cov}(I_i,I_j) \quad (34c)$$
$$\text{cov}(I_i,I_j) = 1 \quad \text{for } i=j$$
$$= 0 \quad \text{for } i\neq j$$

$$e_3 = [\sigma^2(\omega)_R + p\sigma^2(\Delta\omega)]/2 \quad (34d)$$

$$e_4 = [\sigma^2(\omega)_R - p\sigma^2(\Delta\omega)]/2 \quad (34e)$$

$$e_5 = \sigma^2(\Delta\omega)/2. \quad (34f)$$

Experimental Verification of the Mathematical Model

The equation (34) has been verified by $N=10$ series of measurements of the same diffraction profile of a silicon monocrystal. The mean values $h_{i,0}$, the variances $\sigma^2(h_i)$ and the covariances $\text{cov}(h_i,h_j)$ of the measured intensities have been estimated for each of $n=61$ measuring points or for each of $m=1830$ independent pairs of points respectively. The following equations have been used:

Variances of Intensities in the Bond Method

$$\langle h_i \rangle = \frac{1}{N}\sum_{k=1}^{N} h_{i,k}, \quad k=1,2,\ldots,N, \; i = -(n-1)/2,\ldots,(n-1)/2 \quad (35)$$

where $\langle h_i \rangle$ is the estimator of $h_{i,0}$;

$$s^2(h_i) = \frac{1}{N-1}\sum_{k=1}^{N}(h_{i,k}-\langle h_i \rangle)^2 \quad (36)$$

where $s^2(h_i)$ is the estimator of $\sigma^2(h_i)$;

$$c(h_i,h_j) = \frac{1}{N-1}\sum_{k=1}^{N}(h_{i,k}-\langle h_i \rangle)(h_{j,k}-\langle h_j \rangle) \quad (37)$$

where $c(h_i,h_j)$ is the estimator of $\mathrm{cov}(h_i,h_j)$.

The first derivatives $h'(\omega_i)$ have been evaluated from the values $h_{i,0}$.

The estimated values of variances (36) and covariances (37) were next treated by the least-squares method to determine the parameters e_2, e_3, e_4 and e_5 of (34).

A satisfying agreement between the values $\sigma^2(h_i)$ and $\mathrm{cov}(h_i,h_j)$ calculated from the model (34) and the values estimated directly from the experimental data using (36) and (37) has been obtained. The examples of the dependence of the variances and covariances on the sample position i are given in Figures 1 and 2.

Discussion

A question arises, when can we accept equations (1) and (2) or when a more complete model of variances and covariances like (32) and (33) or (34) should be taken into account. The answer this question is not possible without detailed calculations and comparing all of the components in (32) and (33) or (34) for given experimental conditions.

Taking into account the normalization [equations (9) and (10)] we can write (34) in the form:

$$\begin{aligned}C_{i,j} = & e_1 H\sqrt{y_{i,0}y_{j,0}} + e_2 H^2 y_{i,0} y_{j,0} \\ & + 4H^2/\omega_h^2[e_3 + ij/p^2 e_4 - |i-j|e_5]y'(x_{i,0})y'(x_{j,0}).\end{aligned} \quad (38)$$

The first component of (38), connected with the counting statistics, is proportional to the intensity level H, the remaining components—to H^2. Substituting—for convenience—in (38):

$$4e_3/\omega_h^2 = f_3 \quad (39a)$$

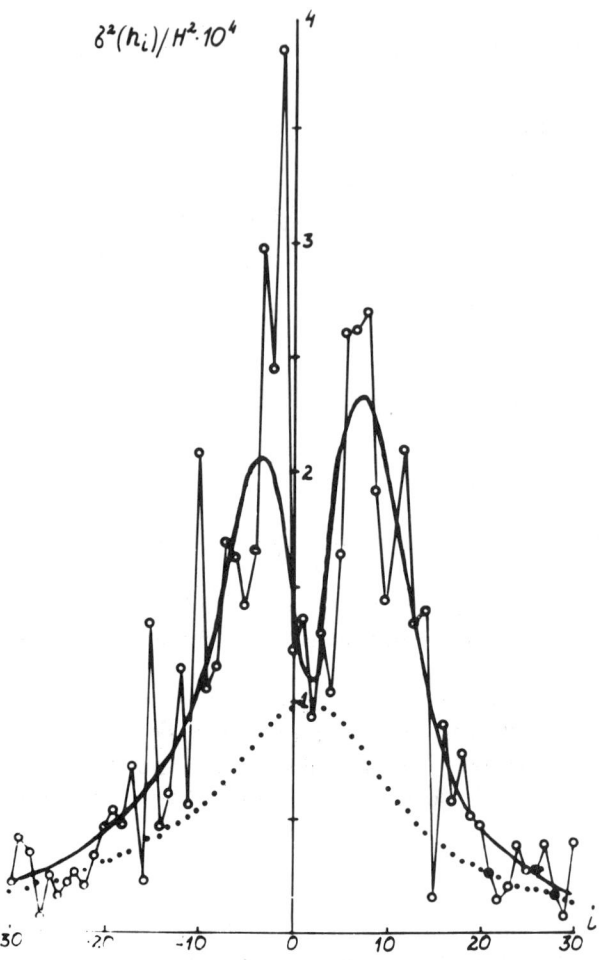

Figure 1. The dependence of normalized values of the variances $\sigma^2(h_i)/H^2$ on the sample position i. ○—○ The values calculated directly from the experiment. —— (continuous line)—The values calculated from the model formula (34) with the parameters e_2, e_3, e_4 and e_5 found by the least-squares method. ····· The values calculated from the formula (1) for the counting (Poisson) statistics only.

$$4e_4/\omega_h^2 = f_4 \tag{39b}$$

$$4e_5/\omega_h^2 = f_5 \tag{39c}$$

and taking H^2 out of the parentheses we obtain

$$C_{i,j} = H^2\{e_1/H\sqrt{y_{i,0}y_{j,0}} + e_2 y_{i,0} y_{j,0} \\ + [f_3 + ij/p^2 f_4 - |i-j|f_5]y'(x_{i,0})y'(x_{j,0})\} \tag{40}$$

As is seen from (40), the relations between the components in (38) depend not only

Variances of Intensities in the Bond Method

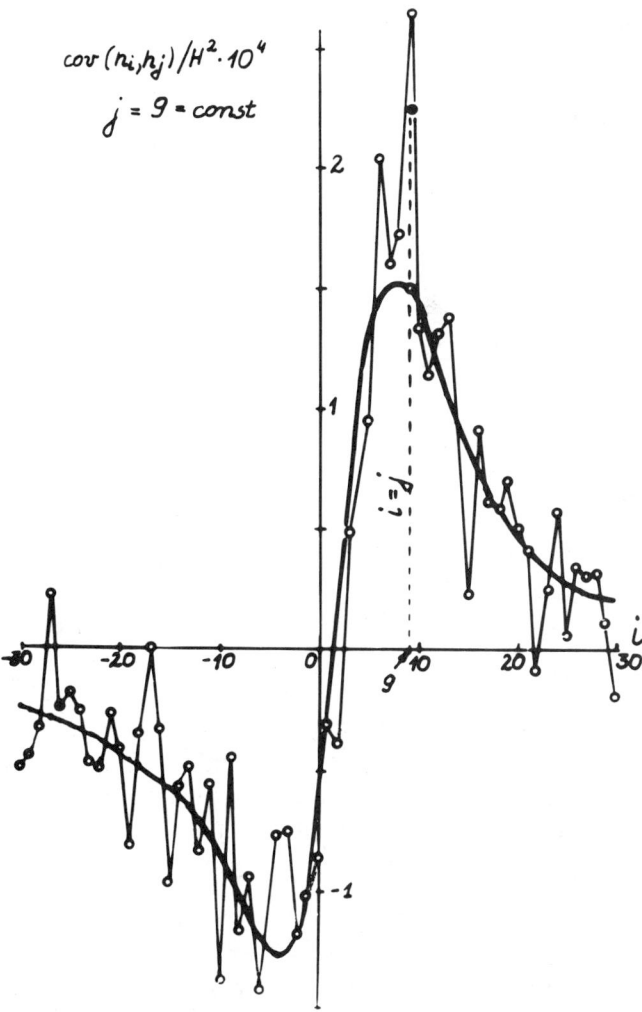

Figure 2. The dependence of normalized values of the covariances $\text{cov}(h_i,h_j)/H^2$ on the sample position i (j=cont. =9). ○—○ The values calculated directly from the experiment. —— (continuous line)—The values calculated from the model formula (34) with the parameters e_2, e_3, e_4 and e_5 found by the least-squares method.

on the values of the parameters e_1, e_2, e_3, e_4, e_5, the shape of the profile and the sample positions $x_{i,0}$, $x_{j,0}$, but also on the level H of the measured intensities. This can be demonstrated by an example.

An example

We assume that the measured profile $h(\omega)$ has a shape of the Cauchy function:

$$y(x)=1/(1+x^2) \tag{41}$$

and its half-with is $\omega_h = 500''$.

The intensities have been measured in $n=21$ sample positions $\omega_i(p=10)$, in the range of angle equal to ω_h, i.e.:

$$\omega_{-p} = \omega_{-10} = -250'', \qquad \omega_p = \omega_{10} = 250''.$$

The values of the parameters characterizing the quality of the experiment are as follows:

$$\sigma(\omega)_R = 2'', \qquad \sigma(\Delta\omega) = 1'', \qquad \sigma(I)/I_0 = 10^{-2}.$$

From the above data we calculate the coefficients e_2, f_3, f_4, f_5 of (40):

$$e_2 = 10^{-4}, \quad \text{for } i=j$$
$$e_2 = 0, \quad \text{for } i \neq j \qquad \text{from (34c)}$$
$$f_3 = 1.12 \times 10^{-4} \qquad \text{from (39a) and (34d)}$$
$$f_4 = -4.8 \times 10^{-5} \qquad \text{from (39b) and (34e)}$$
$$f_5 = 8 \times 10^{-6} \qquad \text{from (39c) and (34f)}$$

Now we estimate the variances of the intensities in selected measuring points—at the edges of the scanning range (ω_{-10} and ω_{10}) and in the centre point ω_0 and the covariances for all the pairs of this points. The normalized values of the angles and intensities and their first derivatives are equal respectively:

$$x_{-10,0} = -1, \quad x_{0,0} = 0, \quad x_{10,0} = 1 \qquad \text{from (9)}$$
$$y_{-10,0} = 0.5, \quad y_{0,0} = 1, \quad y_{10,0} = 0.5 \qquad \text{from (41)}$$
$$y'(x_{-10,0}) = 0.5, \quad y'(x_{0,0}) = 0, \quad y'(x_{10,0}) = -0.5 \qquad \text{from (41).}$$

By substituting the calculated values of the parameters e_2, f_3, f_4, f_5 and the adequate values of $x_{i,0}, y_{i,0}$ and $y'(x_{i,0})$, $(i = -10, 0, 10)$ in (40) we obtain:

$$\sigma^2(h_{-10}) = C_{-10,-10} = 0.5(1/H + 0.82 \times 10^{-4})H^2 \qquad (42a)$$

$$\sigma^2(h_{10}) = \sigma^2(h_{-10}) \qquad (42b)$$

$$\sigma^2(h_0) = C_{0,0} = (1/H + 10^{-4})H^2 \qquad (42c)$$

$$\text{cov}(h_{-10}, h_{10}) = C_{-10,10} = 2.5 \times 10^{-5} H^2 \qquad (43)$$

From the above results and formula it can be concluded that: (a) for low intensities (in above example for $H < 10^3$) the Poisson statistics (the component $1/H$ in the expressions (42a,b,c) for the variances) plays the main role and the influence of the remaining factors (therefore also the covariances) can be neglected; (*b*) for medium intensities ($H \simeq 10^4$) the both components of the variances in the formulas (42*a,b,c*) are significant and the covariances should be taken into account; (*c*) for high intensities ($H > 10^5$), however, the influence of the counting (Poisson) statistics can be omitted and the values of the variances and covariances are comparable.

In the case of diffraction profiles with lower values of ω_h the significance of the counting statistics will be smaller because of the greater values of f_3, f_4 and f_5 (39*a,b,c*).

As results from the above considerations, equations (1) and (2) can be used in very good experiments only for a relatively low intensity level H. However, in the case of "bad" experiment also when dispersions of the measuring parameters are significant, as in the case of a "good quality" crystal, when the intensities are sufficiently large and the half-width of the profile is small (even if the experiment is very good) — the complete models [(32), (33) or (34)] should be taken into account.

Acknowledgements

I thank Professor K. Łukaszewicz for pointing out this subject, Professor A. J. C. Wilson for valuable discussion and Dr D. Kucharczyk and Dr A. Pietraszko for the measurement data.

References and Footnotes

1. E. Urbanowicz, *Acta Cryst.* A*31*, 364-368 (1981).
2. E. Urbanowicz, *Acta Cryst.* A*31*, 369-373 (1981).
3. A. J. C. Wilson, *Acta Cryst. 23,* 888-898 (1967).
4. J. S. Thomsen and F. Y. Yap, *J. Res. Natl Bur. Stand. Sect.* A *72*, 187-205 (1967).
5. A. J. C. Wilson, *Br. J. Appl. Phys. 16,* 665-674 (1965).
6. K. Łukaszewicz, D. Kucharaczyk, M. Malinowski and A. Pietraszko, *Krist. Tech. 13,* 561-567 (1978).

Structure & Statistics in Crystallography, ISBN 0-940030-10-1,
Ed., A. J. C. Wilson F. R. S., Adenine Press, ©Adenine Press 1985.

Precise and Accurate Estimation of Crystallographic Parameters by Maximum-Likelihood and Min-Max Methods

G. B. Mitra, Rabea Ahmed and Prabal Das Gupta
Indian Association for the Cultivation of Science
Calcutta 700 032, India

Abstract

A combination of maximum-likelihood and Min-Max methods has been used in determining extremely precise and accurate values of the following crystallographic parameters:

(i) Lattice parameters of HgClBr and a β-lactam derivative;
(ii) Debye temperature factors of HgClBr and the β-lactam derivative;
(iii) Particle size and strain in pure lead quenched from the melt with intermediate rapidity.

For determining size and strain plots of variance and fourth moment of the line profile *versus* range were used.

The method consists of testing the hypothesis of zero systematic error in the experimental points and then drawing a suitable Min-Max line through them in order to obtain constant bounded errors in the final determinations. The extent of uncorrected systematic error at experimentally observed points and the overall error of measurement were determined for each case. The source of uncorrected systematic error in all cases seemed to be unknown contributions to diffuse scattering in the background. The results have been reviewed also in the light of

(i) the method of maximum correlation;
(ii) the method of moments; and
(iii) the method of minimum chi-squared (χ^2).

It is concluded that a suitable combination of maximum-likelihood and Min-Max methods is the best possible method. An extension of this technique to crystal-structure refinement is suggested.

1. Introduction

All measurements are associated with some errors of different extents. An error is different from a mistake in the sense that mistakes are caused by carelessness on the part of the person making that measurement. An error takes place in spite of the extreme care taken by the experimenter in performing the experiment with the given experimental arrangement and the existing state of theory. An error may also be personal—peculiar to the experimenter, an extreme case being colour-blindness. If the measurement consists of observing a change in colour or a change in the shade of the colour and the observer's retina records a wrong colour or a wrong shade—that introduces an error—however diligent and careful the experimenter may be. This error can be eliminated by having the experiment performed by different experimenters but with identical apparatus and identical methodology. But limitations of instrumentation causes some errors as do the methodology involved. These are called systematic errors. Besides these systematic errors there are errors caused due to unknown and uncontrolled factors. These are called random errors. Better instrumentation and methodology generally helps in reducing these errors but they can not be totally eliminated. Precision of measurement depends on the extent of random errors involved in the measurement. The less the extent of the random error, the more precise the measurement is. It may be stressed again that random error can never be fully eliminated—and hence no measurement can be fully precise, although the degree of precision can be enhanced by various means. A measurement becomes accurate when all errors—systematic as well as random are eliminated. Obviously, no measurement can be made fully accurate—but by removing systematic and random errors by various degrees, a measurement can be made more and more accurate. It may be emphasised that an accurate measurement is automatically precise while a precise measurement need not be necessarily accurate since it might have serious systematic errors.

These considerations are equally valid for measurement of crystallographic parameters. Lattice constants, Debye-Waller factors, average crystallite size and strain in aggregates of distorted crystallities—are some of the parameters which require precise and accurate measurements. There is of course the all important problem of accurate estimation of position parameters as well as thermal parameters of atoms in the asymmetric units in unit cells of crystals. For minimising systematic as well as random errors from these measurements, various experimental as well as analytical techniques have been and are being developed. In this paper, we shall confine ourselves to analytical techniques only.

For quite long time, the analytical technique used for minimising errors from measurements and for estimation of parameters in crystallographic studies has been, invariably, the least-squares method. The least-squares method has some unique advantages. The mathematics is relatively simple and the method can be used for point estimation as well as linear and polynomial function fitting. The least-squares technique, however, cannot identify nor fully correct for systematic errors. (See, however, references (17) and (18), mentioned in §6 below.) Every practitioner of structure determination is aware that until a suitable stage of agreement between calculated and observed structure factor is arrived at, least-squares refinements often lead to unacceptable results. This is probably the stage when the major systematic aberrations have been eliminated and random deviations predominate. But there is no analytical criterion to ascertain when this critical stage has arrived. Because of these and other fundamental reasons to be discussed

Maximum-Likelihood and Min-Max Methods

later, it is desirable to look for some other technique for eliminating errors. Beu, Musil and Whitney (1) used the maximum likelihood method (M.L.M) first introduced by Fisher (2) for estimating the lattice constant of cubic silicon and used the likelihood ratio test (L.R.T.) to examine, identify and determine the level of contribution of different systematic errors. Beu (3) followed this work by determining the lattice constant of cubic tungsten by the M.L.M. while Beu, Musil and Whitney (4) extended these determinations to tetragonal and hexagonal crystals. M.L.M. has not yet been reported to have been used in estimating lattice parameters of crystals of lower symmetry.

Curve fitting as required for Nelson-Riley (5) and Taylor-Sinclair (6) plots for estimating lattice parameters, Wilson (7) plots for determination of Debye-Waller factors, moment-range plots for determination of particle size and strain as developed by Misra and Mitra (8), Mitra and Mukherjee (9) *etc.* have so far been carried out by using least-squares techniques. Because of the known shortcomings of the least-squares method, some other technique seems desirable. M.L.M. has not yet been employed for such purposes in any branch of science. There exists, however, another technique—the Min-Max plot or Chebyshev plot, which is a good candidate for the present purpose and has been used for various purposes although not previously for crystallographic purposes.

A combination of these two techniques—point estimation within the appropriate interval of confidence by M.L.M. and fitting of these points to a linear or polynomial plot by the Min-Max method appear to be a good combination for estimating parameters which are to be obtained by curve fitting. While this is a general technique—and its application is recommended for many fields of study, we propose to describe this for some crystallographic measurements. We shall describe these techniques in greater detail in the following sections.

2. *Maximum-Likelihood and Least-Squares Techniques for point estimation*

Let y_i be the ith observed value of a measurement as a function of an ith independent observable x_i and $f(x_i; a_1, a_2 \ldots a_n)$ be the analytical functional form of y_i in terms of the independent variable x_i and $(a_1, a_2, \ldots a_n)$ are n parameters of measurement which are to be estimated from the measured set $(y_1, y_2 \ldots y_N)$. Then according to the principle of least squares

$$Y = \sum_{i=1}^{N} \omega_i [y_i - f(x_i; a_1, a_2, \ldots\ldots a_n)]^2 \tag{1}$$

will be a minimum. Here the summation is over N observations with $N \geq n$ and ω_i a suitable weight factor whose value may include 1.

This leads to
$$\frac{\partial Y}{\partial a_1} = 0; \frac{\partial Y}{\partial a_2} = 0; \ldots \frac{\partial Y}{\partial a_n} = 0.$$

Solving these n simultaneous equations, the n parameters $a_1, a_2 \ldots a_n$ are determined.

Let $u_i = [y_i - f(x_i; a_1, a_2, \ldots a_n)]$ and let $p(u)$ be the probability density of u. Then the product $\prod_{i=1}^{N} p(u_i)$ is the joint probability of the set of observations $(y_1, y_2 \ldots y_N)$. The natural logarithm of this joint probability

$$L = \ln \prod_{i=1}^{N} p(u_i) = \sum_{i=1}^{N} \ln p(u_i)$$

is called the likelihood of the set of observations i. Since the set i has been actually observed, $\prod p(u_i)$ must be maximum and hence

$$L = \sum_{i=1}^{N} \ln p(u_i)$$

will be maximum. Since $p(u)$ contains the parameters $a_1, a_2, \ldots a_n$, solution of the simultaneous equations $\partial L/\partial a_1 \; 0; \partial L/\partial a_2 = 0; \ldots \partial L/\partial a_n = 0$ will yield the parameters $(a_1, a_2 \ldots a_n)$. If $p(u)$ is a Gaussian density function given by

$$p(u) = \frac{1}{\sqrt{2\pi}\sigma} e^{-x^2/2\sigma^2}$$

we have

$$L = -\frac{N}{2} \ln \pi - \sum_{i=1}^{N} \ln \sigma_i - \frac{1}{2} \sum_{i=1}^{N} \frac{1}{2\sigma_i^2} u_i^2$$

$$= -\frac{N}{2} \ln \pi - \sum_{i=1}^{N} \ln \sigma_i - \frac{1}{2} \sum_{i=1}^{N} \omega_i [y_i - f(a_i; a_1, a_2, \ldots a_n)]^2$$

Here σ_i is the standard deviation of the ith observation and the weight factor $\omega_i = 1/\sigma^2$. It is obvious that the sets of simultaneous equations given by $\partial L/\partial a = 0$ and $\partial Y/\partial a = 0$ are identical and the parameters estimated by these two techniques will also be identical. If, however, $p(u)$ is a function other than Gaussian, the two sets of simultaneous equations will yield two different sets of parameters. The correct set will depend on the proper choice of the probability function $p(u)$. The selection of the proper function $p(u)$ is thus all-important in the correct estimation of the parameters a_i.

It may be mentioned that previous users of M.L.M. in crystallographic parameter estimation like Beu and his collaborators (Beu *et al.* 1962, 1963; Beu 1964) have assumed $p(u)$ to be a Gaussian function. Their estimation of parameter is thus the same as by least-squares method. The superiority and specialty of their work lies in

Maximum-Likelihood and Min-Max Methods

the use of the Likelihood Ratio Test (LRT) in finding out the level of contribution by different types of systematic errors. The only departure is the work of Price (10) who worked out an expression for parameter estimation by M.L.M. using $p(u)$ a Poisson distribution. His expression is not suitable for numerical evaluation of parameters and although he has suggested for structure refinement by M.L.M. with Poisson distribution as the probability density function for intensities nothing as yet has been done in this direction.

Indeed, Gaussian distribution and least-square estimation hold the supreme sway till now. One of the reasons for the popularity of the Gaussian distribution is the relative ease with which the calculation can be carried out. A non-Gaussian distribution, more often than not, lands one in deep mathematical difficulties. However, it has been pointed out by Rothbauer (11) that Gauss (1839) wrote a letter to Bessel dated 26 February of that year in which he opined that the probability of a least-squares solution is infinitely small. Cramér (12) has also said that—'a normal distribution can never be exactly produced by the composition of non-normal components. The composition of a large number of non-normal components produces an approximately normal distribution'. $P(u)$ is composed of a large number of components many of which are sure to be non-normal in nature and so $P(u)$ is rarely normal—although it is asymptotically normal. $P(u)$ is to be fitted to a Pearson curve or the Gram-Charlier or Edgeworth series. As Mood (13) has pointed out that in general $P(u)$ is unknown and all that is available is a series of values of u_i. By means of these values of u_i moments or some other identifier can be computed. By comparing the same moments or the identifying function with similar functions of different Pearson Curves, the theoretical form of $P(u)$ is approximated. This function is then used in estimating parameters by M.L.M. This procedure may involve the user in intractable mathematical difficulties as we shall presently see. As Mood (13) has said—'The theoretical advantage of the principle of Maximum Likelihood over the principle of least squares may become unimportant when it comes to a matter of choosing, say, between a 40 hour and a 10 hour computation. However, it appears desirable to arrive at a $P(u)$ which is at least approximately correct and use this in M.L.M. estimation of parameters rather than use the usual least-squares method.

3. Min-Max and Least-Squares Methods of Curve Fitting

In equation (1), y_i is the ith of N measurements at the point x_i. Hence (y_i, x_i) represents a point and from these data, the different parameters $(a_1, a_2, \ldots a_n)$ are determined so that the best value y at x is finally determined. y determined in this way contains some errors but the errors can be eliminated or their level of significance tested by the L.R.T. technique. This is the technique of point estimation.

Equation (1) is also applicable for curve fitting where x_i's are different points along the x-axis and y_i's are the corresponding measurements along the y-axis. In this case also the parameters $(a_1, a_2, \ldots a_n)$ can be determined by solving the simultaneous equations

$$\frac{\partial Y}{\partial a_1} = 0; \frac{\partial Y}{\partial a_2} = 0; \ldots \frac{\partial Y}{\partial a_n} = 0.$$

In this case also, the basic assumption is that

$$\{y - f(x_i; a_1, a_2, \ldots a_n)\}$$

obeys the Gaussian law of distribution. The distribution may be different as well—say some of the Pearson functions or the Gram-Charlier or Edgeworth series. For such cases, M.L.M. is the appropriate technique. But in such cases, the mathematics becomes too difficult to handle and no one as yet has attempted curve fitting by M.L.M.

There is, however, an alternative method called the Min-Max method or the Chebyshev method. In this method, instead of Y,

$$M = \sum_{i=1}^{N} \omega(x_i) |y_i - f(x_i; a_1, a_2, \ldots a_n)| \qquad (2)$$

is minimised. M is obviously the maximum error and the minimum value of M will be the best fit between y and f. Determination of parameters $(a_1, a_2, \ldots a_n)$ is not possible from the condition that M is minimum since M is not readily differentiable in terms of $(a_1, a_2, \ldots a_n)$. It can, however, be proved that a suitable $f(x; a_1, a_2, \ldots a_n)$ exists so that M is minimum. $f(x; a_1, a_2, \ldots a_n)$ is also unique for the minimum value of M. $f(x; a_1, a_2, \ldots a_n)$ is in general a polynomial called the Chebyshev polynomial but for linear plots f is a linear function. As Ralston (14) has pointed out

'Minimising M has the advantage of giving us a sure bound on the error at any data point. The desirability in some circumstances of having such a sure bound on the error, not just at a discrete set of points but over a whole interval is the motivation behind the minimum-maximum error technique'.

Indeed, if the magnitude of the error is the quantity in which we are directly interested, the minimisation of M would be more desirable than minimisation of Y. While rigorous and detailed techniques of constructing Min-max functions have been discussed by Ralston (15) in a highly sophisticated way, Scheid (16) has described the technique in a simple way specially for linear approximations of the min-max function. Starting from proving the existence and uniqueness of the min-max line or the equal error line through any three given points, Scheid (16) has described the method of its construction and then proceeds on to describe the method of exchanges by which more and more points are included in this line which, of course, gets changed with every exchange of triplet of points. The parameters (a_1 and a_2) of the final linear plot yield the desired informations of the plot like the scale factor and Debye-Waller factor in the Wilson (7) plot.

4. Combination of M.L.M. point estimation with Min-Max linear extrapolation

Since parameter estimation with the help of Min-Max or Chebyshev functions by a straightforward functional operation like differentiation is not possible, the plot must first be obtained and from the constants of the equation to the plot—some at least—of the parameters—can be evaluated. This is particularly so when the Min-Max plot becomes a straight line. Intercepts and slopes of this plot with the coordinate axes yield the necessary informations. The bounding error, however, depends on the dispersion of the experimental points which are to be fitted to the Min-Max plot. In order to obtain maximum possible accuracy, these experimental points should be as much free from errors as possible. As we have already seen, M.L.M. provides an adequate method for this purpose and with the help of L.R.T., it is possible to obtain an estimate with the maximum level of confidence. In the present work, we have combined the two methods of M.L.M. and Min-Max plot so that the points fitted to the Min-Max plot are estimated by M.L.M. and their levels of confidence evaluated by L.R.T.

5. Objective of the present work

The objective of the present work is to apply the technique described in section 4 above for precise and accurate measurement of: (1) lattice constants of a highly absorbing orthorhombic crystal—Mercury chlorobromide and of an ordinarily absorbing monoclinic crystal N−(4-nitrophenyl)−2−oxo−3−chloro−4 phenyl−4′−carbethoxy−azetidine $C_{18}H_{15}N_2O_5Cl$; (2) Determination of isotropic Debye Waller factor for these two compounds; and (3) determination of particle size and strain in quenched lead by the method of variance and fourth moment of X-ray diffraction line profiles. It may be mentioned that this is the first time that the lattice constants of an orthorhombic and a monoclinic crystal have been estimated by the maximum-likelihood method. It is the first time that the Min-Max method is being used in crystallographic problems. It is also the first time, as we shall presently see, that a non-Gaussian distribution has been used in the estimation of lattice parameters.

6. Distribution of lattice parameters – an experimental approach

Beu and his collaborators (1,3,4) while pioneering the application of the maximum Likelihood Method have instinctively accepted the distribution of diffraction angles to be Gaussian. Wilson (17,18), however, reported a non-Gaussian distribution of diffraction angles. Hence it has been considered worth while to investigate the distribution of lattice constants. In course of the present investigations, the distribution of lattice constants rather than of diffraction angles has been studied. This is because in a noncubic crystal, the different unit-cell parameters like *a, b, c etc.* are expected to have the same type of distribution but not necessarily the same distribution parameters. The different moments for the different parameters *a, b, c etc.* may be different. This problem is not confronted for cubic crystals for which $a=b=c$ and the distributions are expected to be same with the same distribution parameter like the standard deviation. For non-cubic crystals, this is not necessarily so. The proper method to study the distribution will be to obtain diffraction patterns of the non-cubic crystal repeatedly, determine the lattice constants for each measurement and finally to obtain the distribution function for each lattice parameter. Obtaining the distribution functions in this way, although highly desirable, may not be practicable. For HgClBr for example both photographic as well as counter methods were used. By using Copper K_a radiations (Peak Voltage -35 KV, tube current 15 ma), the photographic technique with a 5.73 cm radius cylindrical camera and Agfa films, it was possible to record 40 lines with an exposure of 30 hours. A chart run at $\frac{1}{4}°$ per minute using a Philips diffractometer yielded only 16 peaks. An attempt to obtain fixed count data at 1% accuracy level (10,000 counts) at intervals of 0.005° at background level and 0.001° at higher intensities proved extremely time consuming. This and the well known fact that counter diffractometry involved many more sources of error than the photographic technique made it necessary that for lattice-parameter measurement the photographic technique was used. The absorption problem was not so difficult with the β-lactam compound—but the error inherent with the diffractometer were still formidable and much more than those inherent with the photographic technique. Moreover, in any case it was decided to use the photographic technique for HgClBr. Thus, to put all these measurements at the same level, it was decided to study the lattice parameters of the β-lactam compound by the same photographic technique.

Another assumption has been made at this stage. It is that the nature of the distribution function for all lattice constants with the same experimental arrangements will be the same irrespective of the sample. The parameter of the distribution will certainly be different but not the functional nature of the distribution. The distribution may be $f(x; a_1, a_2, ... a_n)$ or $f(x; \beta_1, \beta_2, ... \beta_n)$ but not $\phi(x; \delta_1, \delta_2, ... \delta_n)$. Because of this assumption, it was decided to study the distribution of the lattice constants of Copper with the same photographic experimental arrangements as for HgClBr and the β-lactam compound. Polycrystalline copper—spectroscopically pure supplied by Johnson Mathey & Co.—was used for the purpose, since it takes relatively much shorter time to obtain a powder photograph of copper compared to that of HgClBr or even β-lactam. In all 50 sets of diffraction photographs of the same copper sample with the same experimental arrangements were taken.

Various corrections mentioned by Klug and Alexander (19) including the Nelson-Riley plot were applied and the accuracy of the lattice constants could be taken to be of five decimal places. The observed lattice constants were first arranged in a histogram and then a continuous curve was drawn to represent the distribution function. The curve proved to be symmetrical about a mean position as shown in Fig. 1 and 2 respectively. Fig. 3 represents the frequencies of variation of the a values of copper from the mean value. Fig. 4 shows the plot of $\ln P_i$ (where P_i is the frequency of $x_i = a_i - a$ with a the mean value of the measured lattice parameters and a_i the ith measured value of the lattice parameter a) against x^2. The plot of $\ln P$ against x^2 is observed to be non-linear but drooping. This shows that $P(x)$ is not a Gaussian function. Fig. 5 shows a plot of $d^2/dx^2 \ln P(x)$ against x^2. This plot is very nearly linear with a very small deviation from linearity and with a negative slope. This shows that $P(x)$ can be approximated by

$$\ln P(x) = \ln P_o - \alpha x^2 - \beta x^4$$

where P_o, α and β are constant (independent of x).

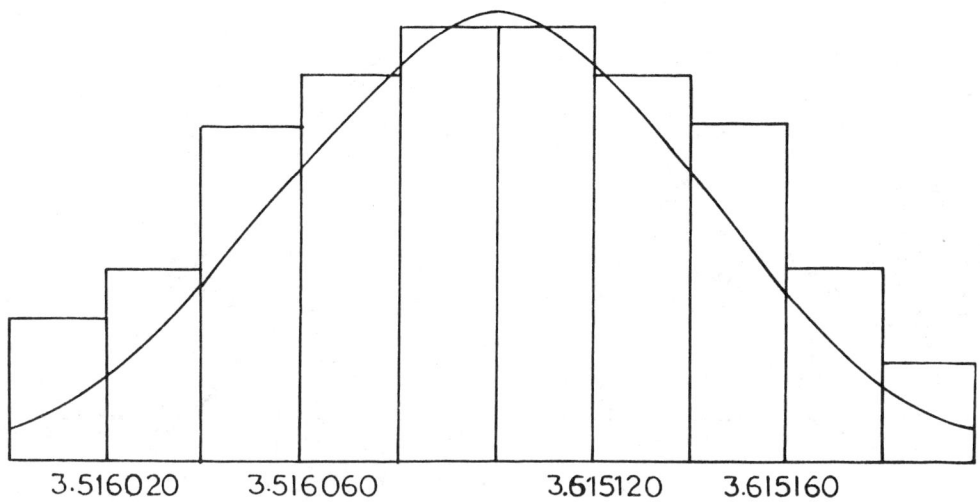

Figure 1. Histogram of the lattice constant of copper.

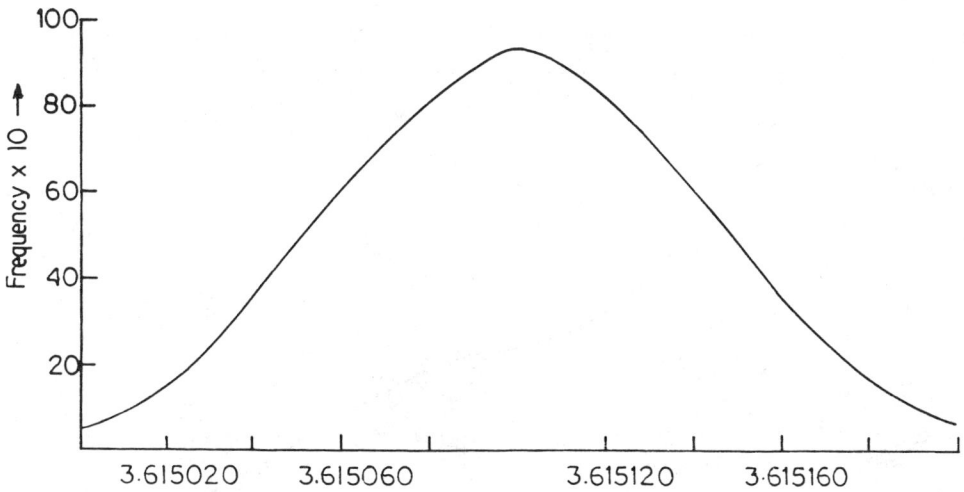

Figure 2. Distribution curve of the lattice constant of copper.

This shows that the distribution $P(x)$ can be fitted to the Pearson curve

$$P(x) = P_o(1 - K^2x^2)^m \qquad (3)$$

where K and m are also constants, so that

$$\ln P(x) = \ln P_o - mK^2x^2 - \tfrac{1}{2}mK^4x^4 \qquad (4)$$

terms with K^6x^6 being negligibly small compared to K^4x^4. For equation (2) and

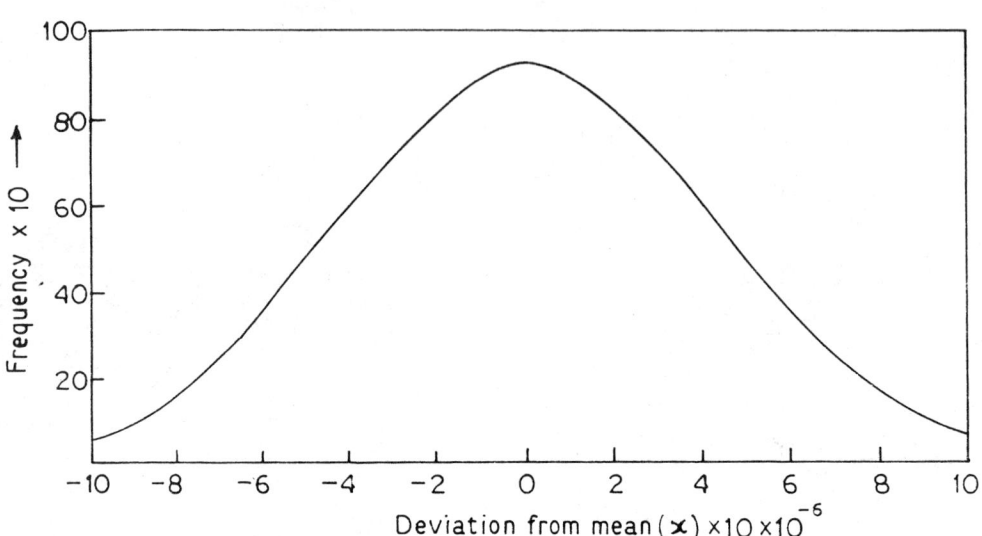

Figure 3. Distribution of the deviation from mean lattice constant for copper.

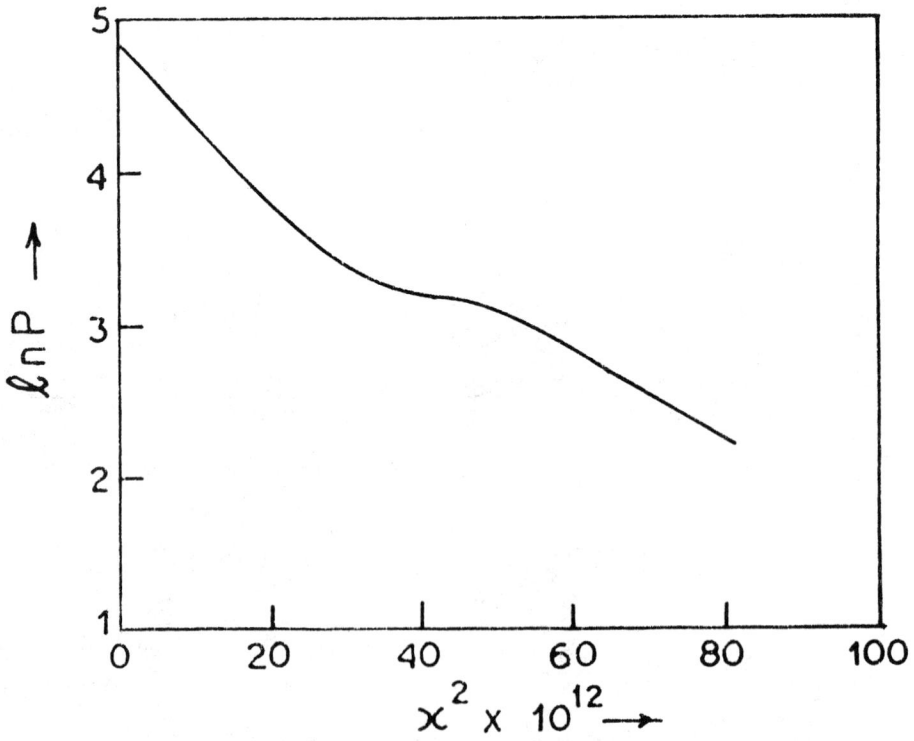

Figure 4. Plot of $\log_e P(x)$ against x^2 for copper, where $P(x)$ is the frequency of x, x_1 is the deviation of the ith measured value of the lattice constant a_i from the mean value a, $x_i = a_i - a$.

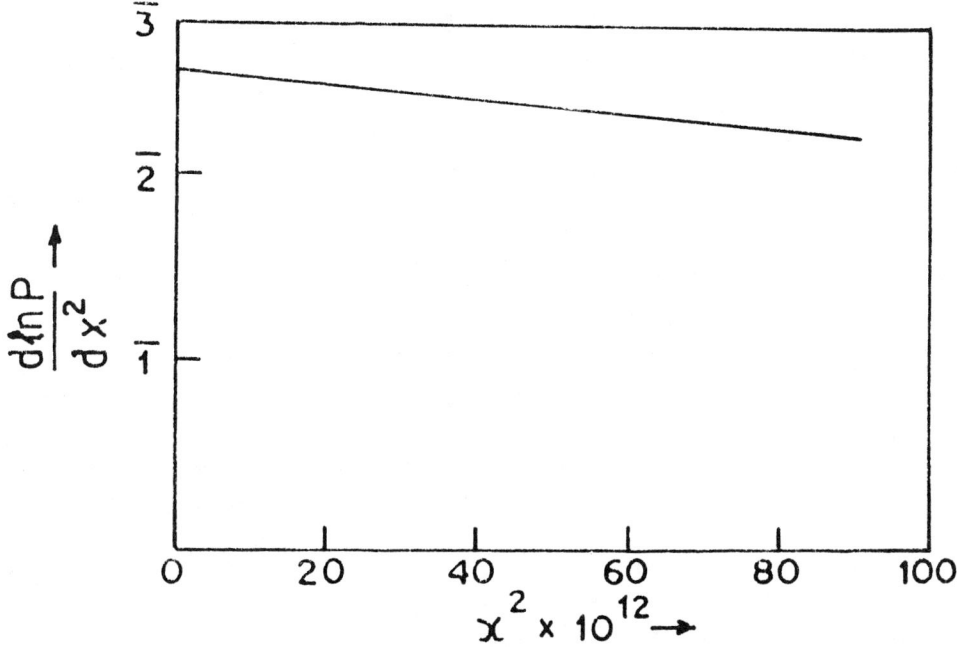

Figure 5. Plot of $\frac{d \log_e P(x)}{d x^2}$ against x^2 for copper.

equation (3) to be consistent, equation (4) has to be quickly convergent so that $K^4x^4 \ll K^2x^2$. Thus, in the first approximation,

$$\ln P(x) = \ln P_o - mK^2x^2 \qquad (5)$$

so that $P(x)$ is very nearly normal—a result which is expected.

7. Estimation of parameter of the distribution function by M.L.M.

Thus, while the ideal distribution function is the Pearson Curve II given by equation (3) the experimental data seem to fit better with its approximate form

$$P(x) = P_o \exp[-mK^2x^2 - \tfrac{1}{2}mK^4x^4]. \qquad (6)$$

We shall now proceed to derive expressions for the parameters P_o, m and K in terms of the observables x_i. The mean lattice constant 'a' has also to be evaluated. Expressions for P_o, m and K for the Pearson Curve II have been given by Elderton (20) but those are in terms of second and fourth moments in terms of observational quantities x_i. While they are appropriate for the ideal distribution given by equation (3), they may not be so for the real distribution applicable to our data. Moreover, our data are few and the moments obtained from them may not be representative of the ideal distribution. From the standpoint of the maximum likelihood method, the distribution given by satisfying equation (4) is more appropriate. From these stand points we start to evaluate the estimate the parameters, P_o, m and K, afresh.

Equation (6) can be written as

$$P_o(x) = P_o \, e^{-K^2 x^2} (1 - \tfrac{1}{2} m K^4 x^4)$$

This, in combination with $\int_0^\infty P(x)dx = \tfrac{1}{2}$ leads to

$$P_o = \frac{K\sqrt{m/\pi}}{1 - 3/8m}. \tag{7}$$

If there are N observables $(x_1, x_2, \ldots x_i \ldots x_N)$, we can write the Maximum likelihood function

$$L = N \ln P_o - mK^2 \sum_i x_i^2 - \tfrac{1}{2} mK^4 \sum_i x_i^4.$$

Substituting the expression for P_o from equation (7), we have

$$L = \tfrac{1}{2} N \ln m + N \ln K - \tfrac{1}{2} N \ln \pi - N \ln(1 - 3/8m) \\ - mK^2 \sum_i x_i^2 - \tfrac{1}{2} mK^4 \sum_i x_i^4. \tag{8}$$

The condition $\partial L/\partial m = 0$, yields

$$\frac{1}{2m} - \frac{3/8m^2}{1 - 3/8m} = K^2 \mu_2 + \tfrac{1}{2} K^4 \mu_4 \tag{9}$$

where $\mu_2 = 1/N \sum_i x_i^2$ and $\mu_4 = 1/N \sum_i x_i^4$.

Again, the condition $\partial L/\partial K = 0$, leads to

$$\frac{1}{2m} = K^2 \mu_2 + K^4 \mu_4. \tag{10}$$

Substituting equation (10) in equation (9), we have

$$K^2 = \frac{4}{27} \frac{(\beta_2 - 3)}{\beta_2 \mu_2} \tag{11}$$

where $\beta_2 = \mu_4/\mu_2^2$.

With the help of equations (10) and (11), we can easily estimate m with

Maximum-Likelihood and Min-Max Methods 163

$$m = \frac{1}{\frac{8}{27} \frac{(\beta_2-3)}{\beta_2} \left\{1 + \frac{4}{27}(\beta_2-3)\right\}}.$$

When $\ln(1 - 3/8m)$ can be approximated by $-3/8m$, we obtain a simpler relation for K^2 given by

$$K^2 = \frac{-\mu_2 + \sqrt{\frac{1}{3}\mu_4}}{\mu_4}. \tag{13}$$

Substituting equation (13) in equation (9), we have

$$m = \frac{3}{2} \frac{\mu_4}{\mu_4 - 3\mu_2\sqrt{\frac{1}{3}\mu_4}}. \tag{14}$$

Substituting equation (10) in equation (8) and differentiating the resulting expression for L with respect to a, we have from $\partial L/\partial a = 0$,

$$a = 1/N \sum_i a_i$$

where a is the mean value of a_i's

8. An alternative distribution and estimation of its parameters

An alternative distribution which immediately suggests itself is the Pearson curve No. VII given by

$$P(x) = P_o(1 + K^2 x^2)^{-m}. \tag{15}$$

Proceeding in the same way as in section (7), we obtain

$$P_o = \frac{K\sqrt{m/\pi}}{1 + 3/8m}. \tag{16}$$

The likelihood function is given by

$$L = \tfrac{1}{2} N \ln m + N \ln K - \tfrac{1}{2} N \ln \pi - N \ln(1 + 3/8m) \\ - mK^2 \sum_i x_i^2 + \tfrac{1}{2} mK^4 \sum_i x_i^4.$$

From $\partial L/\partial m = 0$, we obtain

$$\frac{1}{2m} + \frac{3/8m^2}{1 + 3/8m} = K^2\mu_2 - \tfrac{1}{2} K^4\mu^4. \tag{17}$$

From $\partial L/\partial K = 0$, we obtain

$$\frac{1}{2m} = K^2\mu_2 - K^4\mu_4. \tag{18}$$

Substituting equation (18) in equation (17) we obtain

$$k^2 = \frac{\mu_2}{\mu_4}\left\{2\left(2 + \frac{1}{4\mu_4}\right) + \sqrt{\frac{4}{\mu_4^2} + \frac{4}{\mu_4} + \frac{25}{3}\frac{\mu_4}{\mu_2^2} - 9}\right\} \tag{19}$$

and a first approximation from $\ln(1 + 3/8m) = 3/8m$

$$K^2 = \mu_2/\mu_4\,(1 - \tfrac{1}{3}\mu_4/\mu_2) \tag{20}$$

$$m = \frac{1}{2\,\mu_2^2/\mu_4\,(1 - \sqrt{\tfrac{1}{3}\mu_4/\mu_2^2})}. \tag{21}$$

Substituting equation (18) in the expression for L, and from $\partial L/\partial a = 0$, we have

$$a = 1/N \sum_i a_i.$$

9. Discussion of the applicability of the two distributions

For distribution given by equation (3), we obtain from equation (13) and (14)

$$K^2 = \frac{\sqrt{\tfrac{1}{3}\beta_2} - 1}{\mu_2\beta_2} \tag{22}$$

and

$$m = \tfrac{3}{2}\frac{\beta_2}{\beta_2 - \sqrt{3\beta_2}}. \tag{23}$$

Similarly, values of K^2 and m for the distribution given by equation (15) are as follows

$$K^2 = \frac{1 - \sqrt{\tfrac{1}{3}\beta^2}}{\mu_2\beta_2} \tag{24}$$

and

$$m = \tfrac{1}{2}\frac{\beta_2}{1 - \sqrt{\tfrac{1}{3}\beta_2}}. \tag{25}$$

For equation (22) and (23), values of K^2 and m are positive for $\beta_2 > 3$ while for equations (24) and (25), the same will hold for $\beta_2 < 3$. Hence a preliminary determination of β_2 will indicate which of the distributions given by equations (3)

Maximum-Likelihood and Min-Max Methods 165

and (15) respectively will be valid. Incidentally, this finding is just opposite to the criterion given by Pearson and Hartley (21) and Elderton (22), according to whom for curves of type $P(x) = P_o(1 - K^2x^2)^m, \beta_2 < 3$ and for $P(x) = P_o(1 + K^2x^2)^{-m}, \beta_2 > 3$. This discrepancy may be attributed to the approximations to these functions which give rise to really new distributions having new parameters with new properties.

10. Estimation of the lattice spacing

In the expression

$$P(x) = P_o(1 - K^2x^2)^m$$

K^2x^2 has to be less than 1 since for all values of m, $P(x)$ has to be positive. Thus Kx is also less than 1 and hence may be written as

$$Kx = \sin\theta,$$

so that

$$x = (1/K) \sin\theta$$

and

$$dx = 1/K \cos\theta d\theta.$$

Thus

$$P = P_o \cos^{2m}\theta$$

and

$$\langle x \rangle, \int_{-\infty}^{+\infty} xP(x)dx = 2P_o \int_0^n 1/K \sin\theta \cos^{2m}\theta \, (1/K) \cos\theta d\theta = \frac{2P_o}{K^2(m+1)}.$$

Substituting for P_o from equation (7), we have

$$\langle x \rangle = \frac{2\sqrt{m/\pi}}{K(m+1)(1 - 3/8m)}. \tag{27}$$

Thus, the expectation value of a

$$\langle a \rangle = a + \langle x \rangle = a + \frac{2\sqrt{m/\pi}}{K(m+1)(1 - 3/8m)}.$$

Putting $\frac{d^2P(x)}{dx^2} = 0$, we have, as shown by Scarborough (23)

$$x^2 = \frac{1}{K^2(2m-1)}, \tag{28}$$

which is the mean-square deviation or

$$1/N \sum_i a_i^2 = a^2 + \frac{1}{K^2(2m-1)}.$$

Thus, the R.M.S. value of a is given by

$$a_{RMS} = \sqrt{a^2 + \frac{1}{K^2(2m+1)}}. \tag{29}$$

Similarly, for the distribution $P(x) = P_o(1 + K^2x^2)^{-m}$, starting with $Kx = \tan\theta$ one obtains

$$\langle x \rangle = \frac{2\sqrt{m/\pi}}{K(m-1)(1+3/8m)} \tag{28a}$$

and

$$\langle x^2 \rangle = \frac{1}{K^2(2m+1)}. \tag{29a}$$

11. Derivation of an expression for L.R.T.

The likelihood ratio λ_{LR} is defined by

$$\ln \lambda_{LR} = L(\omega) - L(\Omega)$$

where Ω denotes the domain of the parameter space and ω the domain of the subspace under hypothesis. As assumed by Beu, Musil and Whitney (1) we also accept for Ω, $\sum_i x_i = 0$ and for ω all x_i's are individually zero provided that x_i's denote the systematic parts of the errors.

From $\partial L/\partial m = 0$, where L is given by equation (8), we have

$$mK^2(\mu^2 + \tfrac{1}{2}K^2\mu_4) = \tfrac{1}{2} - \frac{3/8m}{1 - 3/8m} \tag{30}$$

Similarly from $\partial L/\partial K = 0$, we have

$$\frac{1}{2m} = K^2(\mu_2 + K^2\mu_4). \tag{31}$$

Maximum-Likelihood and Min-Max Methods

From equations (30) and (31), we have

$$\ln(mK^2) = \ln\left(1 - \frac{3/4m}{1 - 3/8m}\right) - \ln\left(\mu_2 + \frac{1}{2mK^2}\right). \tag{32}$$

Substituting equations (30), (31) and (32) in the expression for L, we have

$$-2L = N\left[\ln \pi - \ln\left(1 - \frac{3/4m}{1 - 3/8m}\right) + 2\ln(1 - 3/8m) + 1 - \frac{3/4m}{1 - 3/8m}\right] + \sum_i n_i \ln\left(\mu_2 + \frac{1}{mK^2}\right) \tag{33}$$

where n_i is the number of measurements on reflection i and ia labels the ath measurement on the ith reflection. Obviously $\sum_i n_i = N$.

Now $\mu_{2i} = \dfrac{\sum_i (a_{ia} - a)^2}{n_i}$ where a_{ia} is the value of the ath measurement of a from the ith line, or

$$\mu_{2i} = \sum_{ia} \frac{\{(a_{ia} - a_i) - (a_i - a)\}^2}{n_i}$$

$$= \sum_{ia} \frac{(a_{ia} - a_i)^2}{n_i} + 2(a_i - a)\sum_{ia}\frac{(a_{ia} - a_i)}{n_i} + \sum_i \frac{(a_i - a)^2}{n_i}.$$

Let us put

$$S_i^2 = \sum_{ia} \frac{(a_{ia} - a_i)^2}{n_i},$$

and since

$$a_i = \frac{\sum_{ia} a_{ia}}{n_i}$$

we have

$$\mu_{2i} = S_i^2 + (a_i - a)^2. \tag{34}$$

Obviously, S_i^2 represents the standard deviations for the random errors while $(a_i - a)^2$ represents the systematic errors. Since in Ω the sum of the systematic errors has been taken to be zero, for Ω, $\mu_{2i} = S_i^2$ while for ω, $\mu_{2i} = S_i^2 + (a_i - a)^2$ where $(a_i - a)^2$ represents a particular type of systematic error. The hypothesis that is to be tested is if $(a_i - a)^2$ is also zero. The level of confidence of the satisfaction of this hypothesis is indicated by the χ^2 test of λ_{LR} or of $\ln \lambda_{\text{LR}}$ or even of $-2 \ln \lambda_{\text{LR}}$

Thus, from equation (34), we obtain

$$-2L(\omega) = \sum_i n_i \ln\{S_i^2 + (a_i-a)^2 + \frac{1}{2mK^2}\}$$
$$+ N\left[\ln \pi - \ln\left(1 - \frac{3/4m}{1 - 3/8m}\right) + 2\ln(1 - 3/8m)\right]$$

and

$$-2L(\Omega) = \sum_i n_i \ln\{S_i^2 + \frac{1}{2mK^2}\}$$
$$+ N\left[\ln \pi - \ln\left(1 - \frac{3/4m}{1 - 3/8m}\right) + 2\ln(1 - 3/8m)\right],$$

so that

$$-2 \ln \lambda_{LR} = [-2L(\omega)] - [-2L(\Omega)]$$
$$= \sum_i n_i \ln\left[1 + \frac{(a_i-a)^2}{S_i^2 + 1/2mK^2}\right] = W(a). \qquad (35)$$

Since for Ω the systematic errors have been assumed to be zero, this compares well with the expression for Gaussian distribution *viz.*,

$$W(a) = \sum_i n_i \ln\left[1 + \frac{(a_i-a)^2}{S_i^2}\right]$$

derived by Beu, Musil and Whitney (1). similar procedure leads to a similar expression for $W(a)$ for the distribution $P(x) = P_o(1 + K^2x^2)^{-m}$

12. Method of Application of Min-Max technique for linear function fitting of discrete points

The method adopted in this investigation for fitting experimental points to a straight line plot is that advocated by Scheid (16). The method is as follows:

From among the points to be fitted to a linear plot, choose a triplet of points (x_1, y_1), (x_2, y_2), (x_3, y_3) and calculate

$$a_1 = x_3 - x_2$$
$$a_2 = x_3 - x_1$$
$$a_3 = x_2 - x_1$$

and also

$$h = -\frac{a_1 y_1 + a_2 y_2 + a_3 y_3}{a_1 + a_2 + a_3}.$$

Then $(x_1, y_1 - h)$, $(x_2, y_2 + h)$ and $(x_3, y_3 - h)$ will be collinear and points (x_1, y), (x_2, y_2) and (x_3, y_3) will be away from this line by equal magnitudes and alternative signs $-h$, $+h$ and $-h$. The slope and the intercept of this equal-error line will be $\frac{y_1-y_2-2h}{x_1-x_2}$ and $y_3 - h - \frac{y_1-y_2-2h}{x_1-x_2}$ respectively. Next calculate h_n, the distance by which the nth point is away from this line for all points n. If all are less than h, then the search is over and this line is the min-max line. If some other point has an error $H > h$, then choose the point with the maximum H and reject one of the points in the original triplet. Then proceed with this new triplet as before and continue the search till the error H in the final triplet is larger than all other errors h for all the points. When the number of triplets to be fitted is large there will be a large number of triplets and a computer is needed to carry on the search. When the points to be fitted is small, manual computation with a desk calculator is quite easy—specially when one becomes experienced and can instinctively pick up the appropriate triplets so that generally not more than four or five cycles of search are needed.

13. *Application of the Min-Max method to the extrapolation of lattice constants from Sinclair-Taylor or Nelson-Riley Plots*

In course of this investigation, the lattice constants were determined by powder photography as described in Section 6. The d-values obtained in each set of observations were corrected for all known sources of errors and then the respective a values were calculated. These observed a values were then plotted against $f(\theta) = \frac{1}{2}(\cos^2\theta/\sin\theta + \cos^2\theta/\theta)$, where θ is the Bragg angle for the d value from which the a value has been determined. A linear plot has been fitted to a and $f(\theta)$ and the intercepts of this plot with $f(\theta)=0$ gives the correct value of a from which the histogram shown in figure 1 has been drawn. Figure 6 shows the evaluation of a for one such set. It shows that line fitting has been carried out by both least-squares and Min-Max methods. The two methods yield two different sets of results. The maximum error shown in figure 6 obtained by the Min-Max method is ± 0.00009 A.U. for the whole set while the maximum error is the least-squares plot ($+0.00012$ A.U.) is considerably larger than this. This, as will be seen in later sections, is valid for all such sets observed during this present investigation. The case for Min-Max plot is thus established beyond all doubt.

14. *Summary of results for the lattice constant of copper*

The observed values for the lattice constant of copper lead to the following values:

Mean $a = 3.615106$ A.U. $1/2mK^2 = .18682 \times 10^{-6}$
$\mu_2 = 1808 \times 10^{-10}$ $P_o = 931.4385$
$\mu_4 = 1042 \times 10^{-16}$ $\langle x \rangle = 3.40206 \times 10^{-5}$
$\beta_2 = 3.2066$ $\sqrt{\langle x^2 \rangle} = 4.31652 \times 10^{-5}$.
$K^2 = 5.2940 \times 10^4$
$m = 50.7258$

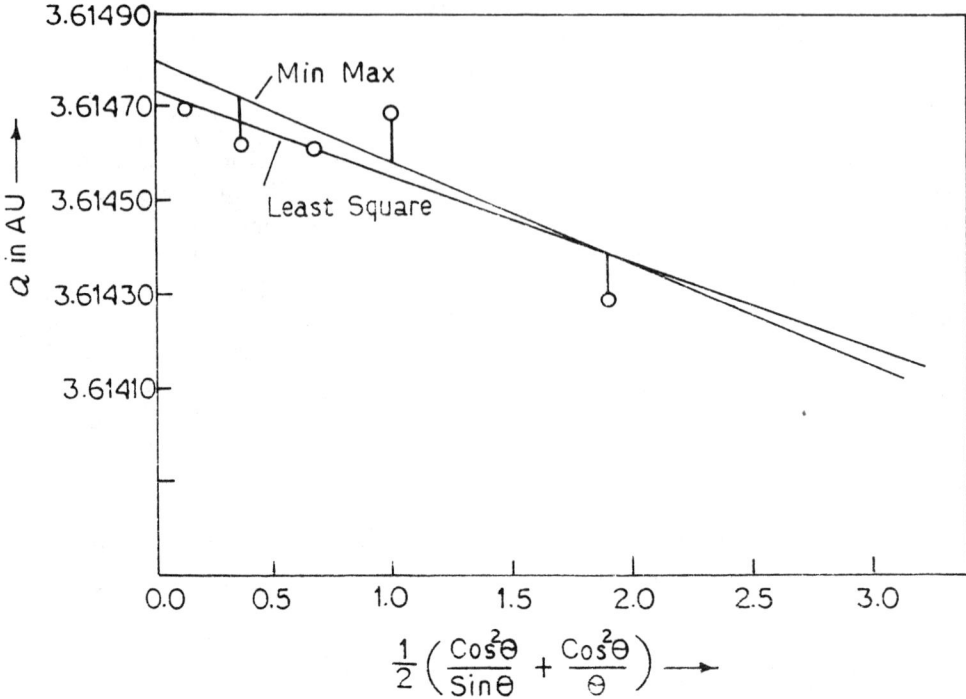

Figure 6. Nelson-Riley plot for copper $f(\theta) = \frac{1}{2}\{(\cos^2\theta/\sin\theta) + (\cos^2\theta/\theta)\}$ against a, where θ is the Bragg angle corresponding to a given reflexion.

The Gaussian approximation would yield

$$\langle x \rangle = \sqrt{2\mu_2/\pi} = 33.8795 \times 10^{-5}$$

and

$$\sqrt{\langle x^2 \rangle} = \sqrt{\mu_2} = 42.4617 \times 10^{-5}.$$

Thus, the Gaussian distribution results in predicting a larger extent of error than the correct distribution. Refraction correction and L.R.T. test for this correction are shown in figures 7 and 8 respectively.

15. Lattice Constants and Debye-Waller Factors for orthorhombic HgClBr and a Monoclinic β-lactam Compound

From the previous investigation, it has been proved that the distribution function for lattice constants with the experimental set up mentioned in Section 6 above is non-Gaussian—given by equation (3). Methods of estimating these parameters by the maximum-likelihood method and the resulting expressions have been described in Section 7 and the expressions for accuracy using these parameters have been

Maximum-Likelihood and Min-Max Methods

given in Section 10. Application of this technique to the estimation of the lattice constant of copper has been described in Section 14. Now, we proceed to apply this method to the determination of the accurate lattice parameter of HgClBr, a newly prepared compound, which belongs to the orthorhombic crystal family and of a β-lactam compound—N—(4-nitrophenyl)—2—oxo—3—chloro—4—phenyl—4'—carbethoxy—azetidine $C_{18}H_{15}N_2O_5Cl$ which belongs to the monoclinic crystal family. Lattice parameters of cyrstals belonging to these two crystal families have not yet been estimated by the M.L.M. Preliminary crystallographic data regarding these two compounds are as follows:

HgClBr

$a = 11.930$ A.U.
$b = 6.669$ A.U.
$c = 10.045$ A.U.

Space group = $P2_122$
Number of molecules per unit cell = 8
Density measured = 5.30 gm cm^{-3}

β-lactam Compound

$a = 10.559$ A.U.
$b = 15.410$ A.U.
$c = 11.313$ A.U.
$\beta = 109.62°$

Space group = $P2_1/a$
Number of molecules per unit cell = 4
Density measured = 1.43 gm cm^{-3}

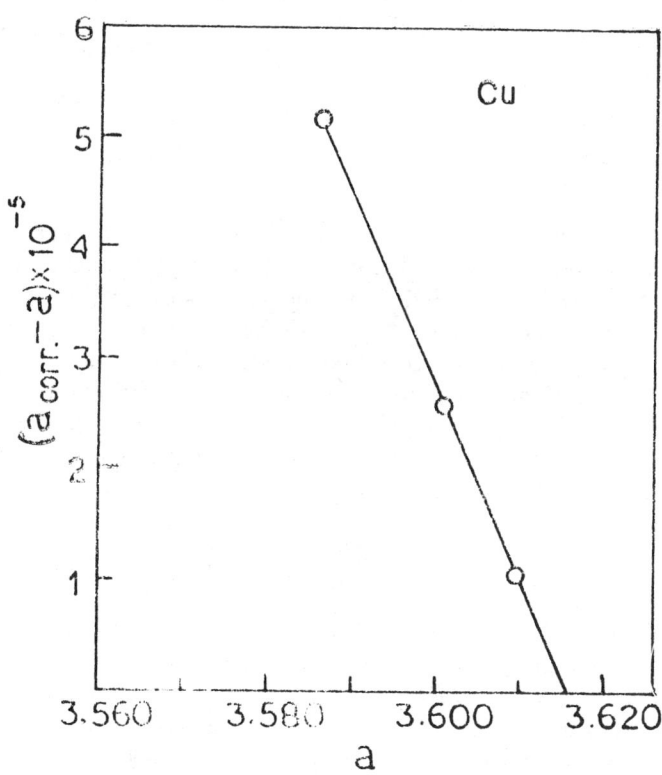

Figure 7. Refraction correction for copper: plot of $a_{corr.} - a$ against a.

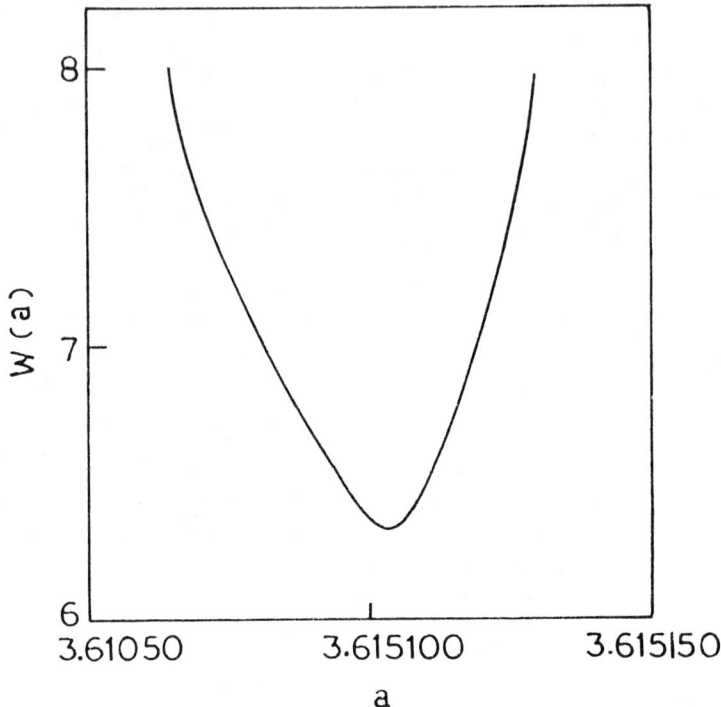

Figure 8. L.R.T. Test for refraction correction of copper.

HgClBr is a highly absorbing material—so it was not possible to prepare a cylindrical sample of so small a radius as to make absorption negligibly small. However, the sample could be made of such diameter that the microphotographs of the different powder lines became symmetrical. All these showed that in spite of the high absorption, the centroid of the line could be considered to be in the same position as the peak maximum. Hence Nelson-Riley plots with multiple-order reflexions at large angles have been considered sufficient to yield correct values of a, b, c for HgClBr. Since reflexions of type $(h00)$, $(0k0)$, $(00l)$ were not many, values of a, b, c were calculated from other reflexions like

$$\frac{1}{d_{211}^2} - \frac{1}{d_{210}^2} = \frac{1}{d_{001}^2} = \frac{1}{c^2}$$

from which the value of c was determined. In this way, 17 values of c, 15 values of b and 18 values of a were determined. With the help of these μ_2 and β_2 for each set were calculated from which m, K^2 and $\langle x \rangle$, $\sqrt{\langle x^2 \rangle}$ were evaluated. $W(a)$ plots (where a represents either of a or b or c) were also constructed from these data and extrapolation of a values to $\sum_i x_i = 0$ and to x (refraction) $= 0$ for $\sum_i x_i$ against a plot yielded the correct values of a, b and c. The results are shown in Table I and plot for a values are shown in figures 9 and 10 respectively.

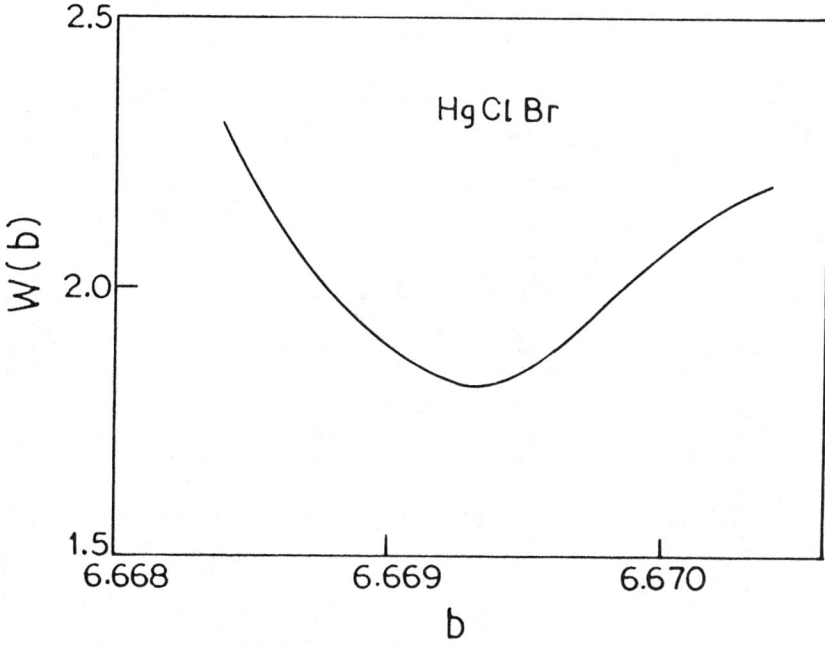

Figure 9. L.R.T. Test for refraction correction for b of HgClBr.

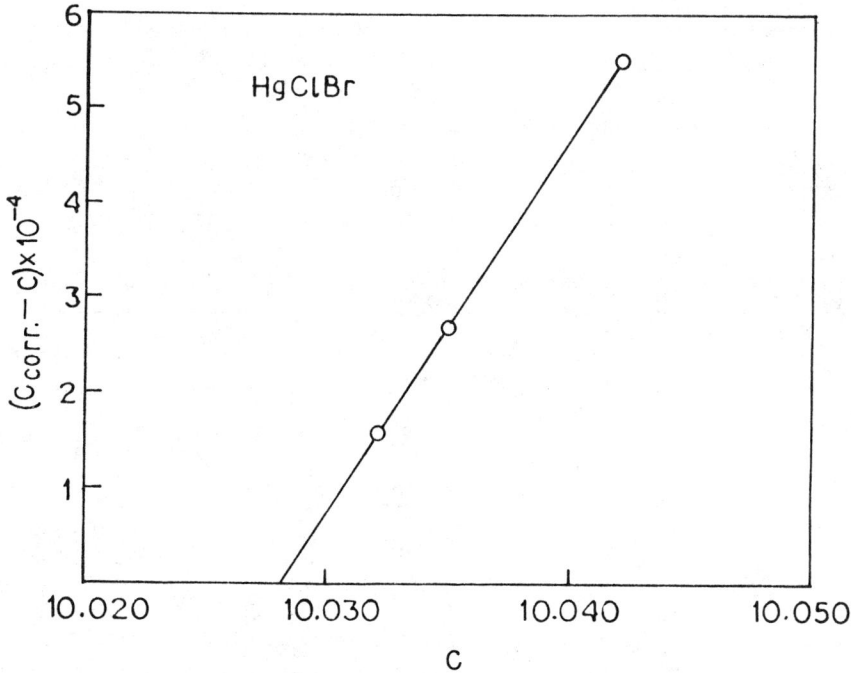

Figure 10. Refraction correction for c of HgClBr.

Table I
Lattice constants of HgClBr

	a	b	c
Final value	11.93026Å	6.66932Å	10.04829Å
μ_2	633×10^{-8}	1222.9×10^{-8}	527.66×10^{-8}
β_2	3.1199	4.8002	4.2203
m	86.2165	7.1052	9.8871
K^2	900.1458	4542.9353	8116.9012
K	30.0024	67.4013	90.0938
$1/2mK^2$	6.442678×10^{-6}	1.54900×10^{-5}	6.2303×10^{-6}
$\langle x \rangle$	1.0053×10^{-3}	2.9062×10^{-3}	1.8799×10^{-3}
$\sqrt{\langle x^2 \rangle}$	2.5456×10^{-3}	4.0819×10^{-3}	2.5617×10^{-3}
P_o	157.85	107.01	166.13
$\langle x \rangle_{GAUSS}$	2.0074×10^{-3}	2.7902×10^{-3}	1.8328×10^{-3}
$\sqrt{\langle x^2 \rangle}_{GAUSS}$	2.5150×10^{-3}	3.4969×10^{-3}	2.2971×10^{-3}

Similar calculations were carried out for the β-lactam compound. Table II shows the relevant parameters and figures 11 and 12 shown the L.R.T. test and refraction corrections for b value of this compound.

16. Determination of Debye-Waller factor for HgClBr and the β-lactam Compound

The Wilson plots $\ln \frac{\langle I \rangle}{\Sigma f_i^2}$ against $\frac{\sin^2\theta}{\lambda^2}$ for the two compounds are shown in figures 13 and 14 respectively. The experimental points have been taken from the Ph.D. thesis of (Mrs) Rabia Ahmed (24) and the D.I.I.T. project report of Shri S. Sen (25). For these two cases M.L.M. was not used in refinement of $\langle I \rangle$. In fact, even if it were applied, it would result in the mean value of $\langle I \rangle$ which has been used here. Assuming I's to be distributed according to the Gaussian Law, the L.R.T. test was carried out and it was observed that refinement of the background level by the

Table II
Lattice constants of β-lactam derivative

	a	b	c	β
Final value	10.55942Å	15.41079Å	11.31342Å	109.6228°
μ_2	1032×10^{-10}	1953×10^{-10}	1034×10^{-10}	2038×10^{-10}
β_2	3.3251	3.4790	3.2830	3.5391
m	32.9427	22.8885	37.5769	20.5211
K^2	140316.34	105522.59	121163.32	1107.13
K	374.5882	324.8424	348.0852	33.2730
$1/2mK^2$	1.0817×10^{-7}	2.0702×10^{-7}	1.0982×10^{-7}	2.2007×10^{-5}
$\langle x \rangle$	2.5762×10^{-4}	3.5363×10^{-4}	2.6015×10^{-4}	3.6355×10^{-3}
$\sqrt{\langle x^2 \rangle}$	3.3142×10^{-4}	4.6004×10^{-4}	3.3362×10^{-4}	4.7494×10^{-3}
P_o	1226.9616	891.4176	1215.9820	86.6234
$\langle x \rangle_{GAUSS}$	25.6318×10^{-5}	35.2607×10^{-5}	25.9036×10^{-5}	20.6680×10^{-4}
$\sqrt{\langle x^2 \rangle}_{GAUSS}$	32.12475×10^{-5}	44.19276×10^{-5}	32.46536×10^{-5}	45.14921×10^{-4}

Maximum-Likelihood and Min-Max Methods 175

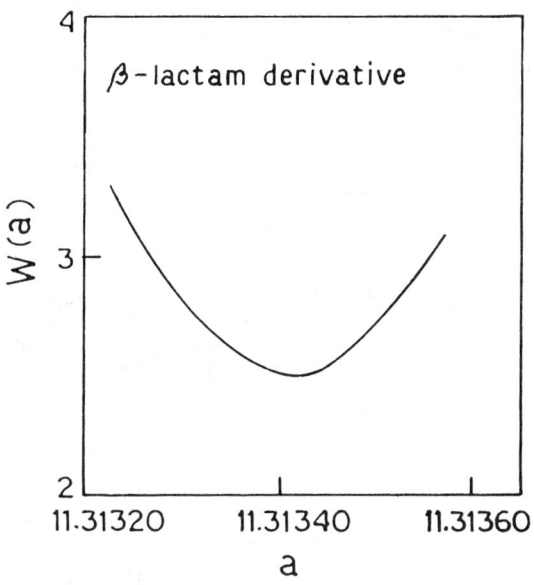

Figure 11. L.R.T. Test for refraction correction for a of β-lactam derivative.

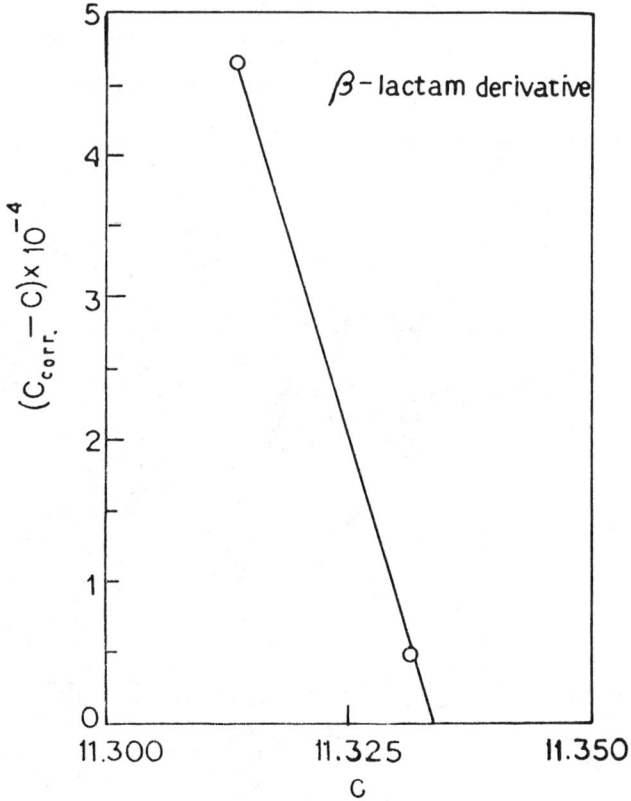

Figure 12. Refraction correction for c of β-lactam derivative.

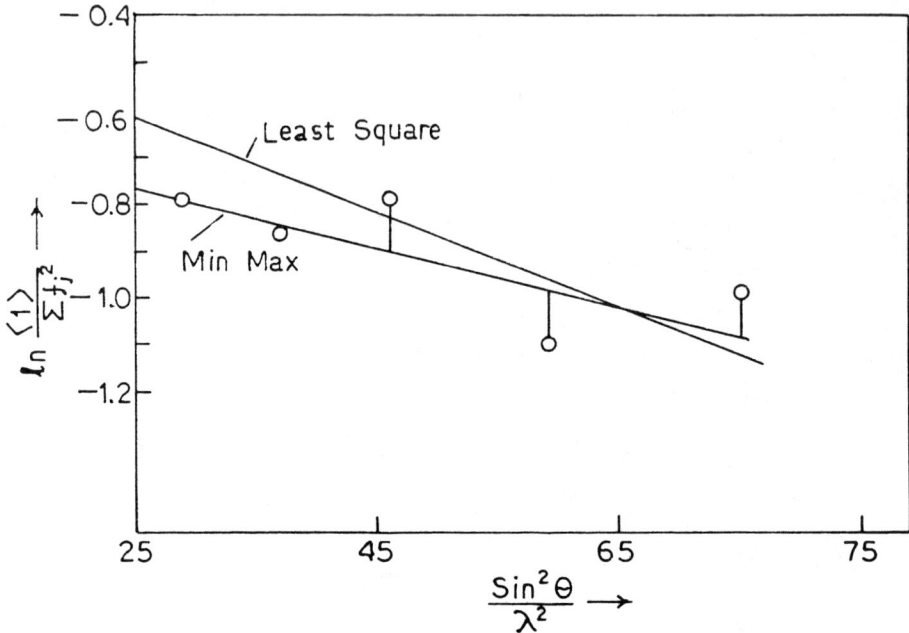

Figure 13. Wilson plot for Debye-Waller factor of HgClBr.

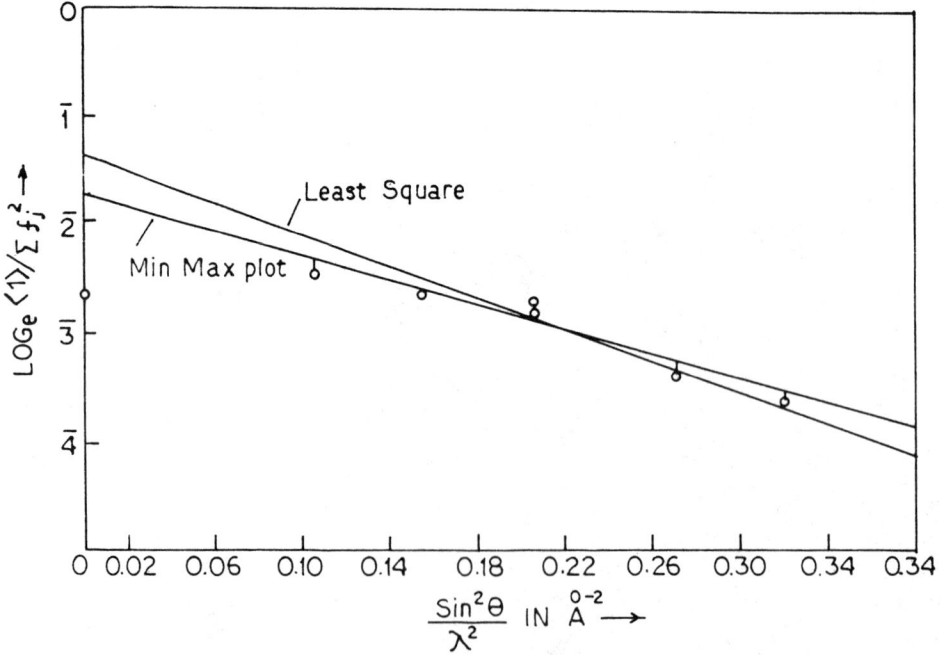

Figure 14. Wilson plot for Debye-Waller factor of β-lactam derivative.

Table III
Measurement of Debye-Waller factors of HgClBr and β-lactam derivative by least-squares and Min-Max methods

Material	Least squares		Min-Max	
	B	θ	B	θ
HgClBr	1.314	90.781	0.819	114.987
β-lactam derivative	4.254	46.846	3.623	50.219

method of Mitra and Misra (26) improved the level of accuracy by a considerable extent. The I's were corrected accordingly. The points were fitted to linear plot both by Min-Max and the least-squares method. These are shown in figures 13 and 14 respectively. For HgClBr, the B value obtained from Min-Max plot agrees much more with that obtained from Lindemann method than the least-squares value does. For the β-lactam compound, the B obtained from Min-Max curve is 3.6, which is surprisingly close to the isotropic thermal parameters finally obtained by Nigam, Ghosh and Mitra (27) for hydrogen atoms. Thus, the Min-Max plot serves a useful purpose.

17. Determination of particle size, strain and strain gradients of rapidly solidified and cold-worked lead

Another important case of linear fitting of experimental points is the determination of particle size and strain in aggregates of distorted crystallities by the single-line technique using the variance of the intensity distribution in the line as a function of the range of observation of the intensity distribution. For such cases, Misra and Mitra (8) showed that the $W\cos\theta/\lambda\sigma$ against λ/σ is linear—the intercept of the line on the $W\cos\theta/\lambda\sigma$ axis indicating the R.M.S. size of the crystallites while the slope of this plot shows the extent of the strains. Here W is the variance of the intensity profile within the range σ and θ and λ are the Bragg angle corresponding to the reflexion and the wavelength used respectively. This is an extremely useful expression since it is valid for even a single order of reflexion. This has been used by Mitra and Chaudhuri (28), Mitra and Mukherjee (9, 29) and many others for various purposes. Similarly Mitra (30) showed that the fourth moment of the line profile is also dependent on the crystallite size and the strain and that the fourth moment due to strain can be separated from the fourth moment due to the crystallite-size profile. He further showed that the fourth moment due to strain as a function of the range of study in reciprocal space against the range is linear, and that from the slope of such linear plot, strain gradients in the aggregate of distorted crystallites can be determined. Linear plots of these two types for lead—cold-worked at room temperature, quenched from melting point at ordinary rate and quenched very rapidly from the molten state have been studied by Ghosh (31) and the present measurements have been taken from his work. Accurate values of crystallite sizes, strains and strain gradients are of very great importance in understanding the mechanism of rapid solidification from melt. Hence, precision of measurement has been increased by the M.L.M. and systematic errors of L.P. correction, background

Table IV
Determination of particle size and strain by the method of variance

Material	Least squares		Min-Max	
	Particle size	Strain	Particle size	Strain
Lead, cold-worked	—	—	12650Å	4.462×10^{-3}
Lead, rapidly solidified	1017.96 A.U.	4.325×10^{-3}	835.42Å	347×10^{-3}

Table V
Determination of strain gradients by the method of fourth moment

e = strain $\qquad t$ = distance in the crystallite

$e' = \dfrac{de}{dt} \qquad e'' = \dfrac{d^2e}{dt^2}$

	Least squares		Min-Max	
	slope $\langle(ee')\rangle$Å^{-1}	Intercept $\langle 4ee''\rangle + \langle 3e'\rangle^2$	slope $\langle(ee')\rangle$Å^{-1}	Intercept $4\langle ee''\rangle + 3\langle e'\rangle^2$Å^{-2}
Lead, cold-worked	−2.625	0.520	−3.052	0.483
Lead, rapidly quenched from melt	−49.960	7.005	−35.762	8.693

determination and T.D.S. corrections have been reduced to 95% confidence level by using the L.R.T. The distribution function of moments is highly complicated but since moments are essentially summations of large numbers of similar quantities, the central-limit theorem has been invoked for this purpose and the distribution has been taken to be Gaussian. Experimental points so estimated are shown in figures 15 and 16 respectively. These points have been fitted to linear plots by the methods of least-squares and Min-Max techniques. The two plots yield different values of intercept and slope. Indeed, as figure 15 shows, the least-squares fitted line for cold-worked lead gives a negative intercept. Since the intercept is a measure of the crystallite size, a negative intercept means a negative magnitude of the crystallite size. This, however, is impossible. The Min-Max plot for the same set of points yields a positive intercept and a positive slope which are as they should be. Hence the Min-Max plot yields useful and accurate results within a maximum boundary of errors.

18. Conclusions and Remarks

The above series of studies establishes that for accurate estimation of crystallographic parameters a combination of the M.L.M. and the Min-Max method is of great help. Since M.L.M. estimation of functions has not yet been possible, curve fitting by M.L.M. has not been attempted. For using the M.L.M., it is necessary to find out the proper probability distribution function which may not be always Gaussian. We have studied the properties of fitting Pearson type II and VII respectively when they can be expanded to the second term only. Experimental studies show that lattice parameters are distributed according to these functions. Secondly the

Maximum-Likelihood and Min-Max Methods

Min-Max linear plot yields better and more definitive results than the least-squares fitted linear plot. In all cases where Min-Max fitting has been used, the corresponding least-squares fits have also been shown—and in all such cases the least-squares plot has at least one point with larger error than the Min-Max plot. Hence use of the Min-Max plot is recommended.

It may be argued that the equal-error Min-Max line is an unique line and is hence deterministic and not probabilistic like the least-squares plot. But the least-squares fit for a set of points is also unique as is the Min-Max fit. When the set of points is modified both the Min-Max and the least-squares fit change. The fact is that the statistics of Min-Max fits have not yet been studied. Probably such studies will be rewarding.

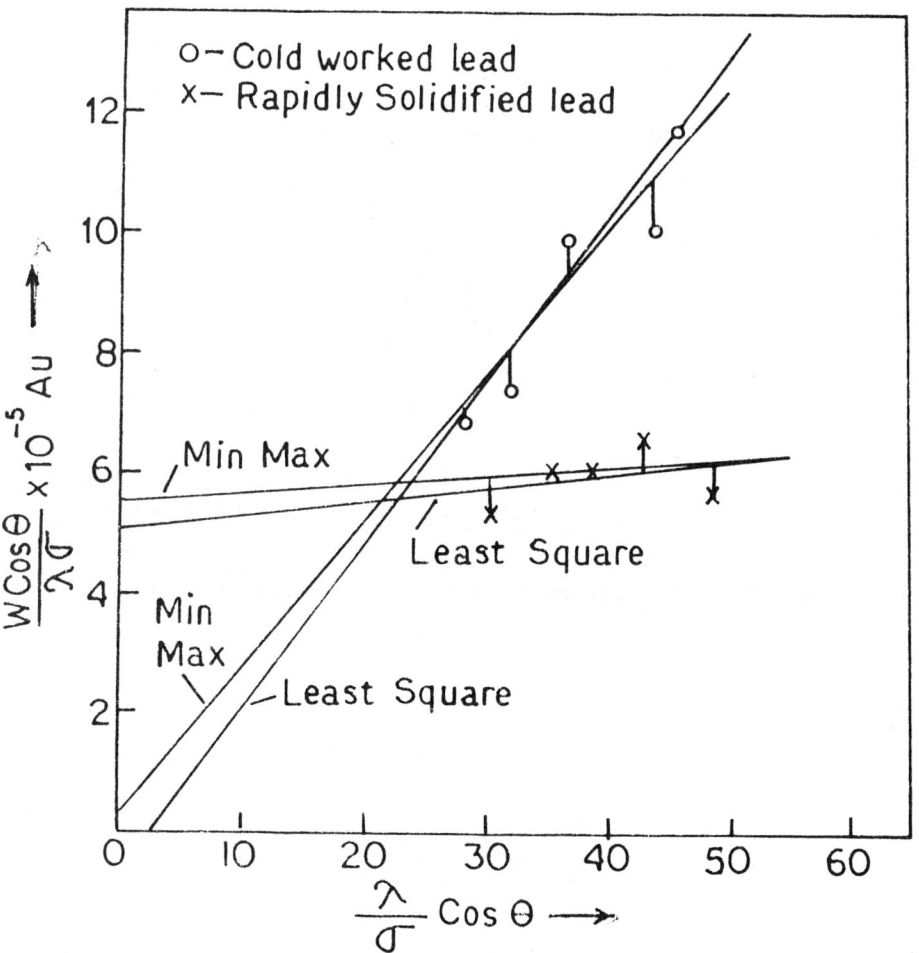

Figure 15. Determination of particle size and strain in lead, cold-worked and rapidly quenched from melt by the method of variance.

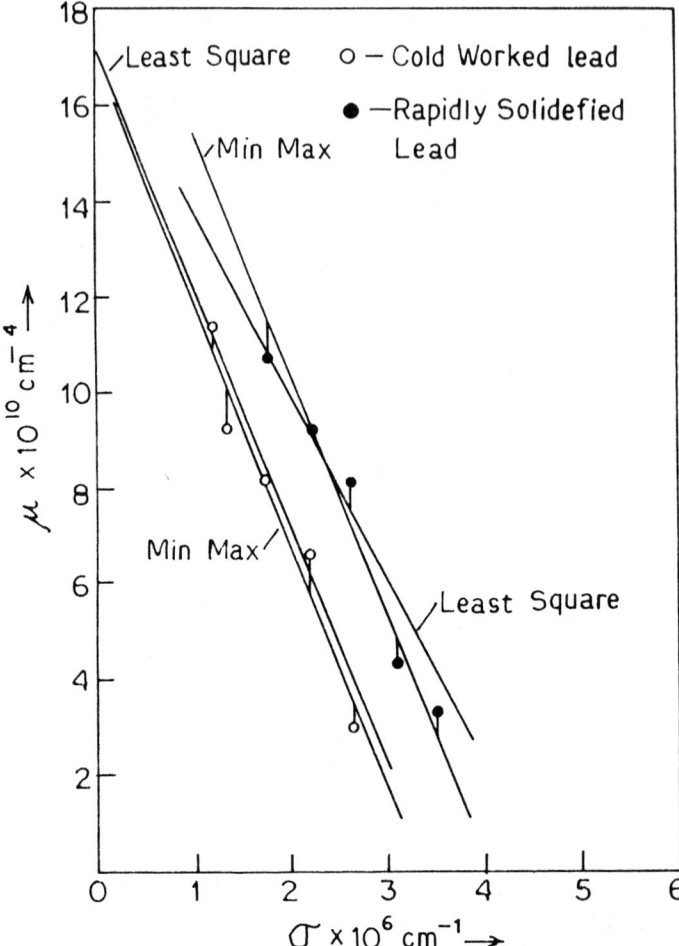

Figure 16. Study of strain gradients in lead quenched and rapidly solidified from melt by the method of fourth moment.

Finally it may be pointed out that the reliability function in crystal-structure determination,

$$R = \frac{\sum_i ||F_o|_i - |F_e|_i|}{\sum_i |F_o|_i}$$

is indeed a maximum error function minimisation, which is a task carried out by crystallographers for crystal-structure determination. Thus extension of the concept and techniques of Min-Max fit may be useful for crystal-structure determination.

Acknowledgements

The authors express their deep sense of gratitude to Professor S.P. Mukherjee, the Head of the Department of Statistics of Calcutta University for his abiding interest in this work and for many illuminating discussions and valuable suggestions. Our heart-felt thanks are also due to Shri Anit Kr. Giri and Miss Paramita Mitra for their help with numerical computational work. Thanks are due to Shri Swapan Kumar Mukherjee for typing the paper and to Shri Porvas Chandra Ghose for drawing the diagrams. One of the authors (G.B.M.) is indebted to the Council of Scientific and Industrial Research for a financial grant enabling him to carry out the work. Thanks are also due to the authorities of the Indian Association for the Cultivation of Science for providing facilities for carrying out this work.

References and Footnotes

1. K.E. Beu, F.J. Musil and D.R. Whitney, *Acta Crystallogr. 15*, 1292 (1962).
2. R.A. Fisher, *Phil Trans. Roy. Soc. 222A*, 309 (1922).
3. K.E. Beu, *Acta Crystallogr. 17*, 1149 (1964).
4. K.E. Beu, F.J. Musil and D.R. Whitney, *Acta Crystallogr. 16*, 1241 (1963).
5. J.B. Nelson and D.P. Riley, *Proc. Phys. Soc.* 57, 160 (1945).
6. A. Taylor and H.B. Sinclair, *Proc. Phys. Soc.* 57, 108 (1945).
7. A.J.C. Wilson, *Nature (Lond.) 150*, 152 (1942).
8. N.K. Misra and G.B. Mitra, *Acta Crystallogr. 23*, 867 (1967).
9. G.B. Mitra and P.S. Mukherjee, *Polymer 21* (2), 1403 (1980).
10. P.F. Price, *Acta Crystallogr. A 35*, (1979).
11. R. Rothbauer, p. 269 in S. Ramaseshan, M.F. Richardson and A.J.C. Wilson (eds.) *Crystallographic Statistics: Progress and Problems.* Bangalore: Indian Academy of Sciences (1982).
12. H. Cramér, *Mathematical Methods of Statistics,* p. 213. Princeton University Press (1957).
13. A.M. Mood, *Introduction to the Theory of Statistics,* p. 120. London: McGraw-Hill (1950).
14. A. Ralston, *A First Course in Numerical Analysis,* p. 231. London: McGraw-Hill (1965).
15. Ref. 14, p. 297.
16. F. Scheid, *Numerical Analysis,* p. 267. Schaum's Outline Series (Asian Student Edn). Singapore: McGraw-Hill (1983).
17. A.J.C. Wilson, p. 360 in *Accuracy in Powder Diffraction.* NBS Special Publication no. 567. Washington DC: US Department of Commerce (1982).
18. A.J.C. Wilson, *Acta Crystallogr. A 36*, 937-944 (1983).
19. H.P. Klug and L.E. Alexander. *X-ray Diffraction Procedures for Polycrystalline and Amorphous Materials.* New York: Wiley (1974).
20. W.P. Elderton, *Frequency Curves and Correlation,* p. 86. Cambridge: University Press (1953).
21. E.S. Pearson and H.O. Hartley, *Biometrika Tables for Statisticians,* Vol. 1, p. 210 (1956).
22. Ref. 20, p. 51.
23. J.B. Scarborough, *Numerical Mathematical Analysis,* p. 498. Calcutta: Oxford and I.A.N. Publishing Co. (1966).
24. R. Ahmed, *Ph.D. Thesis* (unpublished), Aligarh Muslim University.
25. S. Sen, *Diploma of the Indian Institute of Technology Project* (unpublished). Kharagpur: Indian Institute of Technology (1971).
26. G.B. Mitra and N.K. Misra, *Brit. J. Appl. Physics, 17*, 1319 (1966).
27. G.D. Nigam, S. Ghosh and G.B. Mitra, *Cryst. Struct. Communications, 11*, 1672 (1982).
28. G.B. Mitra and A.K. Chaudhuri, *Indian J. Pure and Appl. Physics, 7*, 158 (1969).
29. G.B. Mitra and P.S. Mukherjee, *J. Appl. Crystallogr. 14*, 421 (1981).
30. G.B. Mitra, *Brit. J. Appl. Physics 15*, 917 (1964).
31. T.B. Ghosh, *Ph.D. Thesis* (unpublished). Kharagpur: Indian Institute of Technology (1982).

*Structure & Statistics in Crystallography, ISBN 0-940030-10-1,
Ed., A. J. C. Wilson F. R. S., Adenine Press, ©Adenine Press 1985.*

The Influence of Individual Reflections on the Precision of Parameter Estimates in Least Squares Refinement

E. Prince
Center for Materials Science
National Bureau of Standards
Gaithersburg, MD 20899, U.S.A.

and

W. L. Nicholson
Pacific Northwest Laboratory
Richland, WA 99352, U.S.A.

Abstract

A formula is derived for determining the effect of additional measurements on the precision of refined parameter estimates when a correct structure model has been established. This formula is used in an analysis of the multiple refinements based on the data from the Single Crystal Intensity Project of the International Union of Crystallography, and it is shown that the weighting scheme used in previous studies places a very heavy emphasis on a small number of weak reflections. It is also shown that, if integrated intensities or values of $|F|^2$ are used as the observations, weak reflections have little or no influence on the refinement. An approach to the proper utilization of weak reflections is suggested, and a procedure for improving the precision of parameter estimates when experimental time is limited is proposed.

Introduction

In a previous paper (1) we have described a robust/resistant technique for crystal structure refinement and its use in a reanalysis of the mutiple data sets (2) collected from D(+) tartaric acid in the Single Crystal Intensity Project of the International Union of Crystallography. We confirmed the suggestion of Mackenzie (3) that consideration of extinction would explain some of the variation between data sets in the structures, as refined by Hamilton and Abrahams (4), but, although the robust/resistant technique resolved some problems, the spread of estimated parameter values among the refinements was still greater than would be expected from their computed standard deviations. In the course of our earlier study we observed that, because of the weighting scheme used by Hamilton and Abrahams, certain very weak reflections had a very strong influence on the refinement process.

Some years ago, Shoemaker (5) did a mathematical study of the effect that individual reflections have on the precision of a least squares fit. He was able to arrive at several general conclusions, among which were that position parameter estimates were best optimized by concentrating counting time on high angle reflections, that thermal parameter estimates required low angle reflections also, and that many reflections were not worth measuring at all, except to a minimum level required to establish the correctness and uniqueness of the trial structure. Because of the algebraic complexity of the relations between elements of a design matrix and elements of a variance-covariance matrix, however, he was not able to establish any generally applicable procedure that could be used for optimizing the data collection design for the purpose of precise refinement.

In this paper we apply some recently developed methods of statistical analysis and design (6,7) to the problem of optimizing crystallographic refinement. We first present some general results concerning the effects of adding one more point to a data set that already contains sufficient data to give a non-singular normal equations matrix on the individual elements of the variance-covariance matrix. We then show that the scatter in the values of refined parameters found in D(+) tartaric acid by Hamilton and Abrahams can be accounted for by a high sensitivity of the parameter estimates to small numbers of weak reflections whose standard deviations are underestimated in the weighting scheme they used. Finally, we discuss the implications of this analysis for data collection and refinement procedures and propose an approach to determining an optimum use of measurement time.

Mathematical Analysis

The method of least squares consists of finding the vector $\hat{\mathbf{x}}$, at which the quadratic form $[\mathbf{y} - \mathbf{M}(\mathbf{x})]^T \mathbf{W}[\mathbf{y} - \mathbf{M}(\mathbf{x})]$ is a minimum. Here $\mathbf{y}$ is a vector of experimental observations, and $\mathbf{M}(\mathbf{x})$ represents a set of model functions such that $y_i = M_i(\mathbf{x}) + e_i$, where e_i is an experimental error drawn at random from a population with zero mean, for some unknown $\mathbf{x}$. $\mathbf{W}$ is a weight matrix, which must be positive definite. If the model is the linear function $\mathbf{M}(\mathbf{x}) = \mathbf{A}\mathbf{x}$, the minimum is at $\hat{\mathbf{x}} = (\mathbf{A}^T \mathbf{W} \mathbf{A})^{-1} \mathbf{A}^T \mathbf{W} \mathbf{y}$, so that each element of $\hat{\mathbf{x}}$ is a linear function of the data points, $\mathbf{y}$. If $\mathbf{W}$ is taken equal to $\mathbf{V}_y^{-1}$, where $\mathbf{V}_y$ is the variance-covariance matrix for the joint distribution function of the random errors, e_i, then $\hat{\mathbf{x}}$ is the estimate that has minimum variance among the class of estimates that are linear and unbiased. If $\mathbf{M}(\mathbf{x})$ is not linear, the usual procedure is to find, by numerical methods, a point $\mathbf{x}'$ that is close enough to the minimum for the linear approximation $\mathbf{M}(\mathbf{x}) = \mathbf{M}(\mathbf{x}') + \mathbf{A}(\mathbf{x} - \mathbf{x}')$, where $\mathbf{A}$ is the matrix with elements $A_{ij} = \partial M_i / \partial x_j$, to be a good one. The minimum is then found at $(\hat{\mathbf{x}} - \mathbf{x}') = (\mathbf{A}^T \mathbf{W} \mathbf{A})^{-1} \mathbf{A}^T \mathbf{W}[\mathbf{y} - \mathbf{M}(\mathbf{x}')]$. Although in most practical cases the measurements can be assumed to be independent, and the errors therefore uncorrelated, this is not a necessary condition for what follows, and it introduces no important simplification into the analysis. Because $\mathbf{W}$ is positive definite, there exists an upper triangular matrix $\mathbf{U}$, such that $\mathbf{W} = \mathbf{U}^T \mathbf{U}$. If the errors are uncorrelated, $\mathbf{U}$ is a diagonal matrix with $U_{ii} = 1/\sigma_i$. Designate by $\mathbf{Z}$ the matrix $\mathbf{UA}$ and by $\mathbf{y}'$ the vector $\mathbf{U}[\mathbf{y} - \mathbf{M}(\mathbf{x}')]$. The least squares solution is then $\hat{\mathbf{x}} - \mathbf{x}' = (\mathbf{Z}^T \mathbf{Z})^{-1} \mathbf{Z}^T \mathbf{y}'$.

Influence of Individual Reflections

Let $\hat{\mathbf{y}} = \mathbf{M}(\hat{\mathbf{x}})$, and $\hat{\mathbf{y}}' = \mathbf{U}[\hat{\mathbf{y}} - \mathbf{M}(\mathbf{x}')]$. Then $\hat{\mathbf{y}}' = \mathbf{Z}(\mathbf{Z}^T\mathbf{Z})^{-1}\mathbf{Z}^T\mathbf{y}'$. $\mathbf{Z}$ has dimensions $n \times p$, where n is the number of observations and p is the number of parameters in the model. The matrix $\mathbf{P} = \mathbf{Z}(\mathbf{Z}^T\mathbf{Z})^{-1}\mathbf{Z}^T$ is called the *projection matrix*, because it can be viewed as the projection of an n dimensional vector in observation space onto a p dimensional subspace. $\mathbf{P}$ has dimensions $n \times n$ and rank p. It is evident that $\mathbf{P}^2 = \mathbf{Z}(\mathbf{Z}^T\mathbf{Z})^{-1}\mathbf{Z}^T\mathbf{Z}(\mathbf{Z}^T\mathbf{Z})^{-1}\mathbf{Z}^T = \mathbf{Z}(\mathbf{Z}^T\mathbf{Z})^{-1}\mathbf{Z}^T = \mathbf{P}$, from which many important properties follow. If λ_i is an eigenvalue of $\mathbf{P}$, $\lambda_i^2 = \lambda_i$, so that eigenvalues of $\mathbf{P}$ must be either zero or one, and, because the rank of $\mathbf{P}$ is p, there must be exactly p ones, with the rest zero. Because the trace of $\mathbf{P}$ is the sum of its eigenvalues, the trace is also p. Furthermore, $\mathbf{P}$ is symmetric, so that $P_{ii} = \sum_{k=1}^{n} P_{ik}^2$, and $0 \leq P_{ik}^2 \leq P_{ii} \leq 1$.

A diagonal element of $\mathbf{P}$ represents the rate of change in the calculated value of a data point as a result of a change in the observed value. If P_{ii} is close to one, the model is forced to fit that data point, but the data point has very little influence on the fit at any other data point. If P_{ii} is close to zero, all other elements of the ith row and column must also be close to zero, and therefore that data point has very little influence on the estimate of any parameter or on the fit of any data point. For this reason, the value of P_{ii} is sometimes referred to as the *leverage* of the data point. Because the trace of $\mathbf{P}$ is p, the average value of P_{ii} is p/n, and particularly large values for some data points must be compensated by small values for other data points.

If a data point with small leverage has an observed value markedly different from its calculated value, the fitting algorithm is unable to resolve the discrepancy, and that data point becomes an *outlier*. As Tukey has pointed out (8), if the algorithm is conventional least squares, the effect of trying to fit an outlier is to make the fits of all other data points a little bit worse, with a resulting introduction of bias into the parameter estimates. For this reason robust/resistant procedures (1) can be particularly helpful, because they enable the easy identification of discrepant data points and therefore an assessment of whether the discrepancy results from experimental error or from model inadequacy.

If an observable quantity, y, is known to be a linear function of a control variable, x, the most precise estimates of the two parameters of the function $y = a + bx$ can be made by concentrating measurements at the extreme experimentally accessible values of x. Measurements at intermediate points are needed only to verify that the function is indeed linear. If the projection matrix for this problem is examined, it is observed that the extreme points also have the largest leverage values. We might conjecture that, in a multivariate problem where the model is known to be basically correct, but where the parameters are less precisely known than is desired, there should be a set of exactly p data points that should be measured carefully in order to improve the precision of the parameter estimates most economically. In a non-linear problem like crystallographic structure refinement it is not easy to determine which data points are most important, because the experimentally controllable variables, the reflection indices, are related to the coefficients of the linear approximation, the partial derivatives, by complicated functions. It is nevertheless possible to determine relatively easily the effects of making additional measurements after an approximate model has been established as correct. The development that follows is based on a treatment by Fedorov (7).

Let us suppose that a number, n, of measurements has been made representing a sufficient number of different functions of the parameters for $\mathbf{Z}$ to have full column rank, and therefore for $\mathbf{H}_n = \mathbf{Z}^T\mathbf{Z}$ to be positive definite. The variance-covariance matrix for the parameter estimates after n observations is $\mathbf{V}_n = \mathbf{H}_n^{-1}$. We wish to determine the effect on the precision of the parameter estimates of making one more measurement. We must first define a criterion for evaluating the change in precision. One possible criterion is reduction of the determinant of $\mathbf{V}$, because the determinant, being the product of the eigenvalues, is proportional to the square of the volume of the p dimensional confidence region of the true parameters. Because this is a rather broad criterion, another more focussed one would be to examine the diagonal elements of $\mathbf{V}$, the variances of the marginal distributions of the individual parameter estimates. Before showing how these quantities can be computed, we shall first state a useful lemma from linear algebra.

Lemma 1: Let $\mathbf{A}$ and $\mathbf{B}$ be rectangular matrices with dimensions $j \times k$ and $k \times j$ respectively. Designate by $\mathbf{I}_j$ and $\mathbf{I}_k$ the identity matrices with dimensions $j \times j$ and $k \times k$. Then $(\mathbf{I}_j + \mathbf{AB})^{-1} = \mathbf{I}_j - \mathbf{A}(\mathbf{I}_k + \mathbf{BA})^{-1}\mathbf{B}$. To prove this, postmultiply both sides of the equation by $(\mathbf{I}_j + \mathbf{AB})$, giving

$$\begin{aligned}\mathbf{I}_j &= (\mathbf{I}_j + \mathbf{AB}) - \mathbf{A}(\mathbf{I}_k + \mathbf{BA})^{-1}\mathbf{B}(\mathbf{I}_j + \mathbf{AB}), \\ &= (\mathbf{I}_j + \mathbf{AB}) - \mathbf{A}(\mathbf{I}_k + \mathbf{BA})^{-1}(\mathbf{B} + \mathbf{BAB}), \\ &= (\mathbf{I}_j + \mathbf{AB}) - \mathbf{A}(\mathbf{I}_k + \mathbf{BA})^{-1}(\mathbf{I}_k + \mathbf{BA})\mathbf{B}, \\ &= (\mathbf{I}_j + \mathbf{AB}) - \mathbf{AB}.\end{aligned}$$

Let $\mathbf{z}$ be the row vector with elements $z_j = (\partial M/\partial x_j)/\sigma$, where $M(\mathbf{x})$ is the model function corresponding to an additional measurement, and σ is the standard deviation of the reflection for some fixed measurement time. The new normal equations matrix is, by definition, $\mathbf{H}_{n+1} = \mathbf{H}_n + \mathbf{z}^T\mathbf{z}$, and the new variance-covariance matrix is $\mathbf{V}_{n+1} = \mathbf{H}_{n+1}^{-1}$. $\mathbf{V}_{n+1}$ may be found directly by noting that $\mathbf{V}_{n+1} = (\mathbf{H}_n + \mathbf{z}^T\mathbf{z})^{-1}\mathbf{H}_n\mathbf{V}_n = [\mathbf{V}_n(\mathbf{H}_n + \mathbf{z}^T\mathbf{z})]^{-1}\mathbf{V}_n$, or $\mathbf{V}_{n+1} = (\mathbf{I} + \mathbf{V}_n\mathbf{z}^T\mathbf{z})^{-1}\mathbf{V}_n$. Let $\mathbf{A} = \mathbf{V}_n\mathbf{z}^T$, $\mathbf{B} = \mathbf{z}$, and apply lemma 1, giving $\mathbf{V}_{n+1} = [\mathbf{I} - \mathbf{V}_n\mathbf{z}^T(1 + \mathbf{z}\mathbf{V}_n\mathbf{z}^T)^{-1}\mathbf{z}]\mathbf{V}_n = \mathbf{V}_n - \mathbf{V}_n\mathbf{z}^T\mathbf{z}\mathbf{V}_n/(1 + \mathbf{z}\mathbf{V}_n\mathbf{z}^T)$. A diagonal element of the rank one matrix $\mathbf{V}_n\mathbf{z}^T\mathbf{z}\mathbf{V}_n/(1 + \mathbf{z}\mathbf{V}_n\mathbf{z}^T)$ is therefore the amount by which the addition of this point will reduce the variance of the corresponding parameter estimate. (It should be noted that this analysis applies strictly only to linear models, in which $\mathbf{H}_n$ is independent of $\mathbf{x}'$. In non-linear cases it should be a good approximation provided $\hat{\mathbf{x}}_n$ and $\hat{\mathbf{x}}_{n+1}$ are both close enough to $\mathbf{x}'$ for the linearized model to be an adequate approximation. Numerical methods exist for determining whether the change in $\mathbf{H}$ as a result of the additional step can be neglected or not.)

A particular case of an additional data point is a repetition of one of the points already measured, with the same precision as before. Then, for the ith data point, $\mathbf{z}$ is the ith row of $\mathbf{Z}$, and $\mathbf{z}\mathbf{V}_n\mathbf{z}^T = P_{ii}$, the leverage of that data point. The updated variance-covariance matrix is then $\mathbf{V}_{n+1} = \mathbf{V}_n - \mathbf{V}_n \mathbf{z}^T \mathbf{z} \mathbf{V}_n/(1 + P_{ii})$. Let $\mathbf{t} = \mathbf{z}\mathbf{V}_n$. The quantity $t_j^2/(1 + P_{ii})$ is the amount by which the variance of the estimate of parameter j would be reduced by the repetition of data point i and is therefore a measure of the importance of that data point in the determination of that parameter.

Analysis of the Single Crystal Intensity Project

In the single crystal intensity project of the International Union of Crystallography (2) sixteen laboratories used their usual procedures to collect data sets from crystals of D(+) tartaric acid that had been grown in single batch. One laboratory submitted two data sets, making seventeen in all. Twelve of these experiments contained a sufficient number of reflections to attempt a least squares refinement of the structure. Hamilton and Abrahams (4) performed this refinement, and obtained refined structures for ten of the data sets. In the other two the Gauss-Newton algorithm used by the least squares program failed to find a stable minimum. These authors used a weighting scheme given by $W_i = 1/(0.1|F_i|)^2$, commenting that, "This weighting formula was found to be suitable by an analysis of the average value of $\Delta F/\sigma$ as a function of F. There were few reflections of low enough intensity to be overweighted by this formulation." Comparison of the results of these refinements showed that the spread of values of parameter estimates over the ten experiments greatly exceeded the standard deviations calculated in any experiment.

Subsequently, Mackenzie observed that the differences among the data sets tended to be greatest for the large values of $|F|$, and suggested that this could be explained by differences in secondary extinction in the crystals used in the various experiments. In our earlier study (1) we attempted to reproduce the results of Hamilton and Abrahams and then tested Mackenzie's suggestion by applying an extinction correction of the Zachariasen (9) type in which the absorption-weighted path length, $\overline{T}$, was assumed to be a constant for each experiment and the refined parameter was therefore the product $\overline{T}r^*$. We found that there was a strong correlation between the extinction parameter and the crystal size reported by the experimenters, which is consistent with an extinction effect. The extinction correction also removed some of the disagreement among experiments. Neither extinction corrections nor the application of robust/resistant techniques, however, produced agreement among the experiments that was consistent with their apparent precision.

In the course of these studies it was noticed that the leverages of various reflections were not evenly distributed as a function of $|F|$, but rather that reflections with high values of leverage were disproportionately concentrated among the weak reflections. This led to the speculation that the structure might actually be being determined by a small subset of the reflections whose standard deviations were underestimated in the weighting scheme used by Hamilton and Abrahams, with a resulting underestimate of the standard deviations of the estimated parameters. In order to investigate this possibility, the computer program RFINE4 (10) was modified to write to a disk file all of the elements of the design matrix and all other information necessary to compute the matrix update derived in the previous section.

The structure is monoclinic, space group $P2_1$, with cell constants $a = 7.7290(1)$Å, $b = 6.0004(1)$Å, $c = 6.2126(1)$Å, $\beta = 100.153(2)°$. There are 331 unique reflections with $\sin\theta/\lambda < 0.5$Å^{-1}. The asymmetric unit contains four carbon atoms, six oxygen atoms, and six hydrogen atoms, and the model uses anisotropic temperature factors for carbon and oxygen and isotropic temperature factors for hydrogen. The y parameter of one atom must be held fixed to define the origin, so, including a scale factor and an extinction parameter, there are 115 parameters altogether. Seven experiments (those numbered 2, 4, 5, 7, 8, 9 and 16 by Hamilton and Abrahams) were refined with this model, starting with the parameters of Okaya, Stemple and Kay (11) for carbon and oxygen and parameters given by Cox, Sabine and Taylor (12) for hydrogen. Atomic scattering factors were computed from an analytic formula, with coefficients given by Cromer and Mann (13) for carbon and oxygen and by Forsyth and Wells (14) for hydrogen. This refinement is not quite a repetition of the extinction refinement of our earlier study, because all data sets were merged to unique sets, and all reflections included in each data set were retained throughout the refinement of that set.

After each refinement was completed a computation was made for each reflection of its leverage and of the effect a repetition would have on the variance of each of

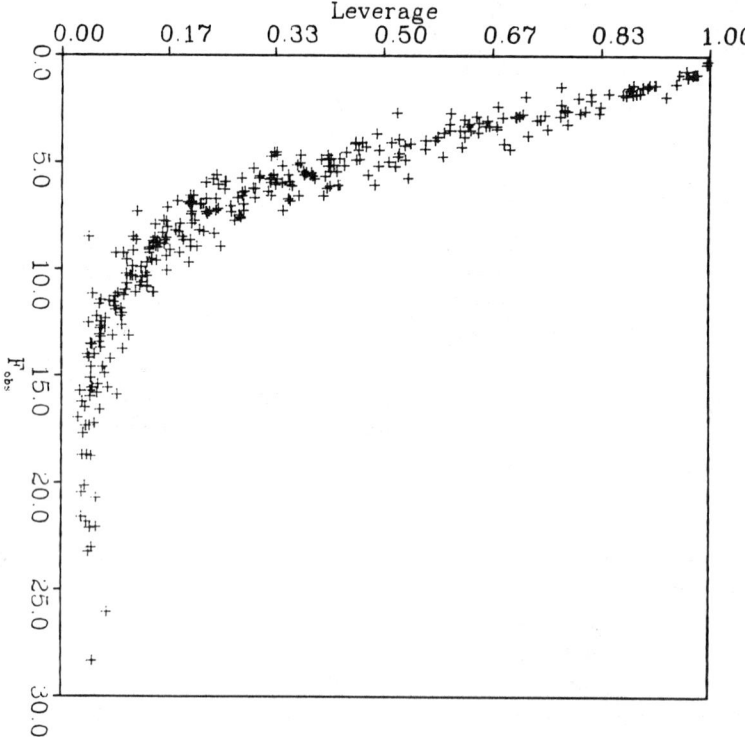

Figure 1. Scatter plot showing the leverage of the reflections in experiment 2, as a function of F_{obs}. The refinement is dominated by the weakest reflections.

Influence of Individual Reflections

the 115 parameters. Figure 1 is a scatter plot in which, for experiment 2, each point represents a reflection with its $|F|$ value plotted on the abscissa and its leverage plotted on the ordinate. It is evident that a few very weak reflections have very high leverage, putting a rather stringent constraint on the corresponding linear combinations of the parameters. For each parameter the five most influential reflections, that is the five reflections that reduced the variance of the estimate of that parameter the most, were determined, and for each reflection the number of times it appears as the most influential, and as one of the five most influential, was determined.

Table I is a summary of information obtained from this analysis. Although there is some variation among the experiments, several things appear consistently in all of the experiments. There is a small number of very weak reflections that has very high leverage (>0.95). However, these reflections do not, in general, also appear among the reflections that are influential in determining particular parameters. There is also a small number of reflections that appear repeatedly, three or more times, as the most influential reflection for several parameters, and, in a few cases, the determination of a parameter is dominated by a single reflection. Except for those determining the scale and extinction parameters, these influential reflections are neither very weak nor very strong, but they tend to be relatively weak. At the other extreme, something more than a third of all reflections, including most of the strong ones, never appear at all in the list of the five most influential reflections for any parameter.

An extreme case of a parameter depending on a single reflection is shown in Table II. In five of the seven experiments the determination of β_{11} for oxygen atom O(3) is dominated by the 711 reflection, usually by more than a factor of two over the next most influential reflection. In the other two experiments this reflection is missing, and the computed standard deviations are substantially larger. Figure 2 shows the value of the parameter, plotted against the value of $|F_o|$, corrected for extinction.

Table I

Summary of refinement information for seven refinements of the structure of D(+) tartaric acid. For each experiment the table gives the experiment number, the total number of reflections, R and R_w, the number of reflections with leverage >0.95, the number of independent reflections that are most influential for determining some parameter, and the number of reflections that do not appear among the five most influential for any parameter.

Exp. number	number of reflections	R	R_w	Leverage >0.95	Most influential	Not in top five
2	332	0.026	0.034	11	78	138
4	330	0.044	0.062	5	76	134
5	326	0.027	0.036	5	78	120
7	331	0.027	0.034	8	78	123
8	320	0.040	0.060	5	71	117
9	302	0.089	0.109	6	85	96
16	325	0.024	0.031	6	75	109

Table II

The value of β_{11} of O3 compared with the value of F_{obs} of the 711 reflection (corrected for extinction). The computed standard deviations of the least significant figures are given in parenthesis. The 711 reflection is missing from data sets 8 and 9.

Experiment number	$\beta_{11(O3)}$	F_{obs}
2	0.00569(46)	7.03
4	0.00635(83)	6.64
5	0.00479(45)	7.22
7	0.00444(43)	7.07
8	0.00567(87)	—
9	0.00378(151)	—
16	0.00364(41)	7.59

The range of parameter values is about seven standard deviations, and the parameter value is strongly correlated with $|F_o|$ for this single reflection.

These results suggest that the scatter in refined parameter values observed by Hamilton and Abrahams is due to a relative overweighting of weak reflections and emphasize that reliable assessments of the uncertainties of estimated parameters in least squares require that data points be weighted in approximate proportion to the reciprocals of the variances of the distribution functions for the observations.

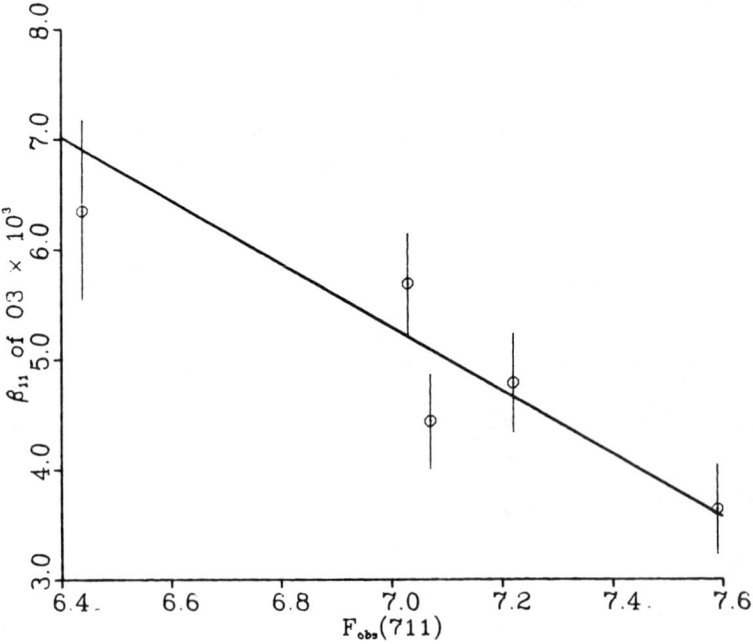

Figure 2. A plot of β_{11} for O3 against F_{obs} of the 711 reflection (corrected for extinction). Error bars are 1σ, as computed in the refinement.

Influence of Individual Reflections

Weak Reflections

A number of authors (15,16) have discussed the problem of the proper procedure for handling very weak reflections, particularly when the subtraction of background from the intensity where a peak is expected yields a negative number, in structure refinement. It is a common practice to omit reflections whose net intensity is less than 2σ, where σ is the standard deviation of the observation as computed from counting statistics, from the refinement, but some authors have suggested that this procedure may introduce a bias into the parameter estimates, and it has also been suggested that the "observations" used in refinement should be either the integrated intensities or $|F|^2$, with the negative values included, even though the true net intensity is obviously a non-negative quantity.

Actually, omitting a weak reflection, or *any* reflection, cannot bias the parameter estimates, as may be shown as follows. We have already seen that least squares assumes that the observations can be represented by $y_i = M_i(\mathbf{x}) + e_i$, where e_i is a random variable with zero mean. If we use angle brackets to indicate expected values, this can be written $\langle \mathbf{y} \rangle = \mathbf{M}(\mathbf{x})$, and in the linear approximation, $\langle \mathbf{y} - \mathbf{M}(\mathbf{x'}) \rangle = \mathbf{A}(\mathbf{x} - \mathbf{x'})$, where $A_{ij} = \partial M_i/\partial x_j$ evaluated at the point $\mathbf{x} = \mathbf{x'}$. The least squares estimator is $\hat{\mathbf{x}} - \mathbf{x'} = (\mathbf{A}^T\mathbf{W}\mathbf{A})^{-1}\mathbf{A}^T\mathbf{W}[\mathbf{y} - \mathbf{M}(\mathbf{x'})]$ so $\langle \hat{\mathbf{x}} - \mathbf{x'} \rangle = \langle (\mathbf{A}^T\mathbf{W}\mathbf{A})^{-1}\mathbf{A}^T\mathbf{W}[\mathbf{y} - \mathbf{M}(\mathbf{x'})] \rangle = (\mathbf{A}^T\mathbf{W}\mathbf{A})^{-1}\mathbf{A}^T\mathbf{W}\langle [\mathbf{y} - \mathbf{M}(\mathbf{x'})] \rangle = \mathbf{x} - \mathbf{x'}$. Therefore, provided the observations are unbiased estimates of $\mathbf{M}(\mathbf{x})$, and the elements of $\mathbf{W}$ are not functions of the errors, the least squares estimator is an unbiased estimator for $\mathbf{x}$ independent of $\mathbf{W}$, and a zero weight for any point does not introduce a bias. Nevertheless, as French and Wilson (17) have pointed out, a zero measurement is as good a measurement as any other, perhaps better, and we should therefore address the question of whether the inclusion of the very weak reflections can contribute to improved precision in the parameter estimates. Analysis of the projection matrix can give an answer to this question.

We have seen that the addition of one more measurement to the data set changes the variance-covariance matrix for the estimated parameters by $\mathbf{V}_{n+1} = \mathbf{V}_n - \mathbf{V}_n\mathbf{z}^T\mathbf{z}\mathbf{V}_n/(1 + \mathbf{z}\mathbf{V}_n\mathbf{z}^T)$, where $\mathbf{z}$ is a row vector with elements $z_j = (\partial M_{n+1}/\partial x_j)/\sigma_{n+1}$. Now, if the net integrated intensity is the observed quantity, $M_{n+1}(\mathbf{x}) = Lp|F_{n+1}(\mathbf{x})|^2$, where L and p are the Lorentz and polarization factors. Then $z_j = Lp[d|F_{n+1}(\mathbf{x})|^2/d|F_{n+1}(\mathbf{x})|][\partial |F_{n+1}(\mathbf{x})|/\partial x_j]/\sigma_{n+1}$, but $d|F_{n+1}(\mathbf{x})|^2/d|F_{n+1}(\mathbf{x})| = 2|F_{n+1}(\mathbf{x})|$ and is equal to zero if $|F_{n+1}(\mathbf{x})|$ is equal to zero, so that $\mathbf{z}$ is null, and $\mathbf{V}_{n+1} = \mathbf{V}_n$. The leverage of the reflection is $\mathbf{z}\mathbf{V}_{n+1}\mathbf{z}^T = 0$, so inclusion of this measurement has no influence whatsoever on the estimates of either the parameters or their standard deviations. Because the integrated intensity is related to $|F|^2$ by the multiplicative constant Lp, the same result applies on refinement with $|F|^2$ taken as the observed quantity.

It remains for us to discuss the question of whether the inclusion of the very weak reflections can improve the precision of estimates if the observed quantity is taken to be $|F|$. An apparent difficulty arises because, when $|F|$ is large enough to be observed, the standard deviation of $|F|$, σ_F, is approximately related to the standard deviation of the integrated intensity, σ_I, by $\sigma_F = \sigma_I/[2|F|(Lp)^{1/2}]$. This expression becomes infinite when $|F|$ approaches zero, which would result in a zero weight for that reflection, but the extrapolation to zero $|F|$ to assign a value to σ_F is obviously not valid. The value of $|F|$ is less than some threshold above which the reflection would certainly be visible above background, and the value of F therefore lies within a circular region in the complex plane, or a line segment on the real axis if the crystal has a center of symmetry, defined by the threshold.

A probability density function for F, from which a mean and variance can be inferred in order to assign a value and a weight, can be constructed with the aid of Bayes's theorem (17,18), which can be stated $\Phi_c(F|I) = Cl(I|F)\Phi_p(F)$, where $\Phi_c(F|I)$ is the conditional distribution function for F given an observed value for the integrated intensity, I, $l(I|F)$ is the likelihood function of observing the value of I given a value of F, and C is a normalizing constant to ensure that the integral over all possible values of F is unity. $\Phi_p(F)$ is called the prior distribution for F and is based on what is previously known. In this case the prior distribution is derived from the partially refined trial structure. If it is assumed that the "correct" value for each element of $\mathbf{x}$ lies within a limited range of the trial structure, the range of F_c also lies within a range that is centered on F_c for the trial structure with a width that is proportional to the square root of its leverage. (The standard deviation of F_o that is used is that which can be achieved in some specific maximum counting time.) If the reflection is observed, the likelihood function has appreciable values in an annular region in the complex plane with a mean radius equal to $|F_o|$, which becomes two peaks centered at $\pm |F_o|$ if there is a center of symmetry. Application of Bayes's theorem results in the phase of F_c being applied to F_o, in agreement with usual crystallographic practice. Two extreme possible situations are illustrated in fig. 3 for the case of unobserved real structure factors. In one situation (Fig. 3a), the reflection has low leverage, and the range of F_c lies entirely within the likelihood function, which is uniform between the positive and negative threshold values. F_o, which is the mean of the distribution shown by the dashed lines, is then equal to F_c, and including the reflection adds nothing to the information. In the other situation (Fig. 3b) the reflection has high leverage, and the prior distribution includes the likelihood function. The distribution is then identical to the likelihood function, the mean of F_o is 0 and $\sigma_F = t\sqrt{3}$, where t is the threshold value. Including an unobserved reflection of this type will have a beneficial effect on the precision of parameter estimates.

It should be noted that this analysis applies only in the final stages of refinement, where the model is known to be essentially correct and the aim is to improve the precision of estimates of parameters in order to establish the details. It is important to have a high level of over-determination, and to pay careful attention to assuring that reflections that are not observable above background also have calculated values that would not be observed, in the initial stage where the details have not been worked out and it is necessary to avoid false minima and to establish the uniqueness of the solution. Although these stages are often referred to as

Influence of Individual Reflections

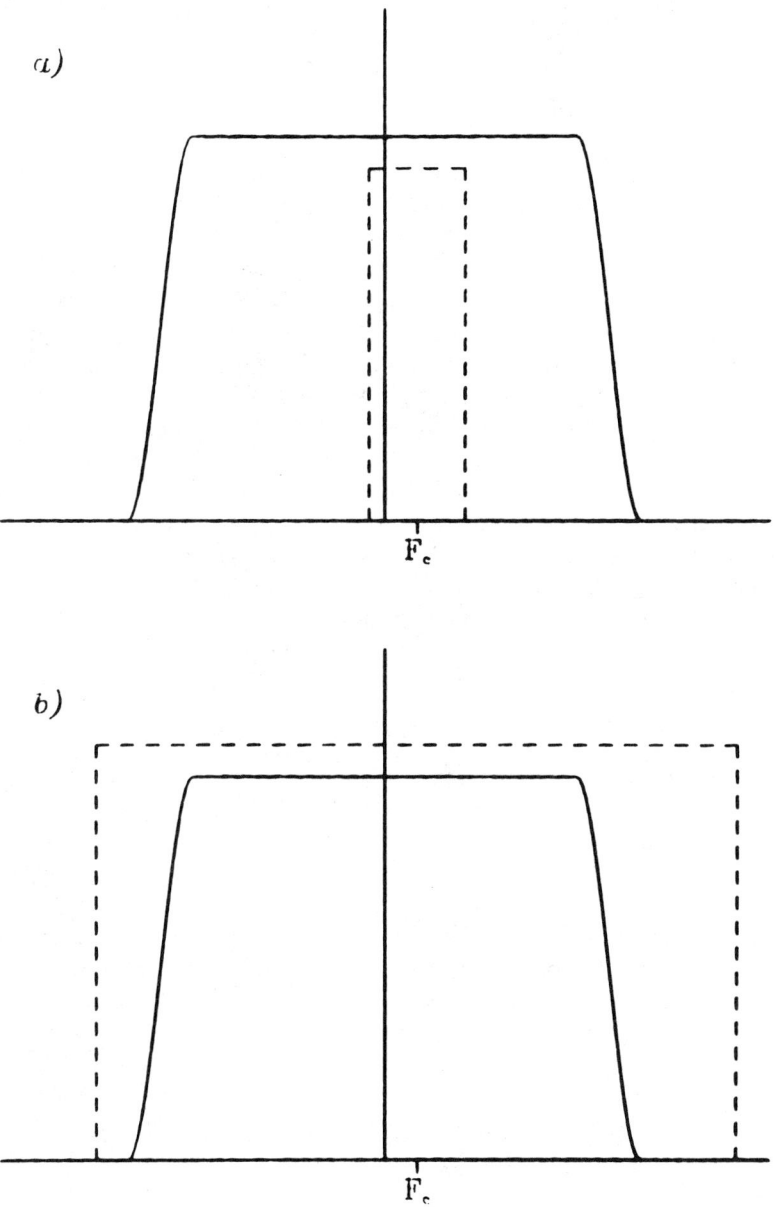

Figure 3. A Bayesian interpretation of the distribution of the value of F_{obs} for an unobserved reflection. In each case the solid curve is the likelihood of not observing the reflection, as a function of F. It is uniform in the region below the threshold of observability. In *a* the reflection has low leverage, and the posterior distribution is identical to the prior (dashed lines). In *b* the reflection has high leverage, and the posterior distribution is identical to the likelihood. The mean of this distribution is $F = 0$.

"refinement", the problems are numerical rather than statistical. The sum of squares is used as a convenient overall figure of merit, and well developed numerical techniques exist for finding its minimum, but statistical analysis is significant only in the near vicinity of the global minimum.

Optimization of Experiments

We now return to the question of how to choose the set of reflections on which measurement time should be concentrated in order to make the most precise estimates of structure parameters. Since the elements of **Z** depend on σ_i, the standard deviations of the observations, as well as on the partial derivatives, it is necessary to define a reasonable value to use for the standard deviation of a measurement that has not yet been made. Suppose that the standard procedure is to count for equal time intervals at n_p points in the region of the Bragg peak and at n_b points in background regions, getting total intensities I_p and I_b in these regions, respectively. Then the magnitude of F is given by $|F| = (1/s)\{[I_p - (n_p/n_b)I_b]/(Lp)\}^{1/2}$, where s is the scale factor, and its standard deviation is given by $\sigma_F = (1/2s)\{[I_p + (n_p/n_b)^2 I_b]/[I_p - (n_p/n_b)I_b]Lp\}^{1/2}$. In the limit of large peak to background ratio the expression for σ_F approaches the value $(1/2s)(Lp)^{-1/2}$, which is dependent on Bragg angle, but independent of $|F|$. If something is known, from preliminary measurements, about the threshold level for observable reflections, these expressions can be used directly to determine approximate values of σ_F, with weak reflections being treated as indicated in the previous section. Otherwise, all reflections can be assigned standard deviations equal to $(Lp)^{-1/2}$, with the realization that this will tend to exaggerate the usefulness of the weakest reflections.

A reasonable experimental procedure would then start with a rapid measurement of all reflections within the limiting sphere in reciprocal space. Because weak reflections carry little information for either Fourier or direct methods for determining the initial model, the procedure should minimize the time spent measuring them. When the initial model has been determined the reflections that have the most influence on each parameter of the model may then be determined, and additional time devoted to precise measurement of the intensities of these reflections, which may include some weak ones whose partial derivatives with respect to some parameters are large.

The results on the tartaric acid data suggest that, if only the most influential reflection for each parameter were to be included, the data set that would be chosen for more precise measurement would be smaller than the number of parameters. This is not necessarily inconsistent, because it is possible that some linear combinations of parameters are actually determined rather precisely by even a rough data set, and measurements that determine other linear combinations may improve the precision of several individual parameters at once. However, it is certainly good practice to include a larger set, perhaps including the two or three most influential reflections for each parameter, because this will reduce the potential sensitivity of the refinement to undetected, nuisance effects that could bias an apparently good refinement.

Influence of Individual Reflections

Conclusions

From this analysis, a number of general conclusions can be drawn. It is possible, by means of straightforward computation, to determine which reflections make the most important contributions to the precision of the refinement. The validity of computed standard deviations depends critically on the use of an appropriate weighting scheme, and rather small subsets of the data, if sufficiently overweighted, can cause serious bias. If refinement is on integrated intensity or on $|F|^2$, weak reflections contribute almost nothing to parameter estimation or to improved precision. If refinement is on $|F|$, a few of the weak reflections may be important. Finally, the precision of a refinement may be enhanced by concentrating measurement time on a relatively small subset of the accessible reflections. Many reflections, including most, but not all, of the weak ones contribute little to improved precision, although they are important to establishing the correctness, and uniqueness, of the model.

References and Footnotes

1. W. L. Nicholson, E. Prince, J. Buchanan and P. Tucker in *Crystallographic Statistics,* Eds., S. Ramaseshan, M. F. Richardson and A. J. C. Wilson, Indian Academy of Sciences, Bangalore, p. 229 (1982).
2. S. C. Abrahams, W. C. Hamilton and A. McL. Mathieson, *Acta Cryst. A26,* 1 (1970).
3. J. K. Mackenzie, *Acta Cryst. A30,* 607 (1974).
4. W. C. Hamilton and S. C. Abrahams, *Acta Cryst. A26,* 18 (1970).
5. D. P. Shoemaker, *Acta Cryst. A24,* 136 (1968).
6. D. C. Hoaglin and R. E. Welsch, *The American Statistician, 32,* 17 (1978).
7. V. V. Fedorov, *Theory of Optimal Experiments,* trans. W. J. Studden and E. M. Klimko, Academic Press, New York and London, (1972).
8. J. W. Tukey in *Critical Evaluation of Chemical and Physical Structural Information,* Eds., D. R. Lide, Jr. and M. A. Paul, National Academy of Sciences, Washington, DC, p.3 (1974).
9. W. H. Zachariasen, *Acta Cryst. A24,* 212 (1968).
10. L. W. Finger and E. Prince, *A System of Fortran IV Computer Programs for Crystallographic Computations,* NBS Tech. Note 854, National Bureau of Standards, Washington (1975).
11. Y. Okaya, N. R. Stemple and M. I. Kay, *Acta Cryst. 21,* 237 (1966).
12. G. W. Cox, T. M. Sabine and G. H. Taylor, unpublished (1966).
13. D. T. Cromer and J. B. Mann, *Acta Cryst. A24,* 321 (1968).
14. J. B. Forsyth and M. Wells, *Acta Cryst. 12,* 412 (1959).
15. F. L. Hirshfeld and D. Rabinovich, *Acta Cryst. A29,* 510 (1973).
16. A. J. C. Wilson, *Acta Cryst. A36,* 929 (1980).
17. S. French and K. Wilson, *Acta Cryst. A34,* 517 (1978).
18. G. E. P. Box and G. C. Tiao, *Bayesian Inference in Statistical Analysis,* Addison-Wesley, Reading, MA (1973).

Structure & Statistics in Crystallography, ISBN 0-940030-10-1,
Ed., A. J. C. Wilson F. R. S., Adenine Presss, ©Adenine Press 1985.

Principles Involved in the Development of Expert Systems for Data Acquisition

H.J. Milledge[†], M.J. Mendelssohn[†], C.M. O'Brien[‡] and G. I. Webb[‡*]

Crystallography Unit, Geology[†] & Statistical Science[‡]
University College London, Gower Street
London WC1E 6BT, U.K.

Abstract

Procedures for incorporating chemical information into least-squares refinements which were described by Mendelssohn, Milledge and Webb (*Acta Cryst. A31,* S232 (1975)) and which involved various assumptions concerning the ways in which heterogeneous data sets should be combined can also lead to the development of "expert systems" of data acquisition. Such systems are particularly relevant to diffractometer data, which are usually collected in a routine manner unrelated to the characteristics of the data set involved, although the time taken to measure a single reflection and its background, of the order of one minute, gives the associated computer controlling the data acquisition ample time to update its current information and modify its future activities.

An attempt is made here to illustrate the basic principles involved in the design of such an expert system in the hope that future generations of crystallographers may come to regard "routine" data collection as having the same relative efficiency as "unconstrained" least-squares refinements of crystal structures.

Introduction

One aspect of crystal structure analysis which appears to have received little attention up to the present time is the possibility of analysing the data *during the actual collection process* in order to see whether the acquisition of more data is likely to be cost-effective, and, if so, whether some additional reflexions should be obtained from a randomised list (*i.e.* the remaining data should be *sampled*) or whether particular reflections should be selected in the light of some specified criteria.

This concept is particularly relevant to computer-controlled diffractometry, where the data are acquired serially—especially when, as is the case with neutron diffraction, beam-time may be very expensive and/or in short supply—because the operational speed of modern computers means that the time taken to reposition the

*Present Address: H.M. Treasury, Treasury Chambers, Parliament Street, London SW1P 3AG.

crystal on a 4-circle diffractometer and measure a reflexion and its background (of the order of a minute) is long enough for a substantial amount of calculation to be undertaken before the crystal has to be reset to measure another reflection.

The possibility of allowing the data collection procedure itself to be modified by the data already collected is particularly attractive in the case of crystal structure solution and refinement because the resolution attainable is linearly dependent on $\sin\theta/\lambda$ (S), whereas the number N of available reflections hkl is proportional to the third power of S, so that as S increases dN/dS increases very rapidly, and even for unassisted least-squares refinements (the word "unconstrained" is avoided because it has a precise meaning not intended here) the over-determination K of the n variable parameters, $K=N/n$ can easily exceed the value 3 usually assumed adequate for SFLS refinements if even a modest increase in resolution is required.

The approach described here had its origin in an attempt by one of us (G.I.W.) (1) in 1973 to incorporate *via* a Bayesian approach to the problem as a whole the practical experience gained during the operation of a Pegasus computer program written by one of us (2) for structure solution and refinement. Using this program it has been found that for small molecules up to about 30 atoms in the asymmetric unit, trial structures derived from stereochemical criteria and packing considerations could be refined by diagonal least squares using the following criteria:

(a) Starting with a small quantity of low-angle data, having essentially correct phases.

(b) Defining an acceptable fraction, f, so that a reflexion was ignored if $F(\text{calc}) < f.F(\text{obs})$, thus improving the phase discrimination.

(c) "Squaring up" the molecule by obtaining axes and moments of inertia of the molecule distorted during the LS cycle, and repositioning the trial molecule or molecular fragment using the inertia origin and orientation.

(d) Increasing S and f in stages until all reflections were included and squaring up was no longer necessary.

(e) Applying various fractional shifts which effectively damped the oscillations of parameters p when the average partial derivative $\partial p/\partial F$ was small in relation to the average value of dF, so that the refinement did not become unstable.

(f) Examining first, second, third and fourth moments of dF to see whether the remaining errors were normally distributed.

(g) Removing outliers (usually arising from mis-indexing of the oscillation data) if $dF > 3\sigma$ in the final stages.

Expert Systems for Data Acquisitions

The new procedure involved adding to the then current UCL Fortran SFLS program routines (See Appendix 1 for the principles involved) enabling:

1) Variances to be specified individually for each parameter being refined.

2) Variances to be specified individually for the restrictions concerning interatomic distances and planarity being imposed.

3) The weights associated with the observed reflexions to be recalculated for each cycle from a formula which involved, *inter alia*, the experimentally determined weight and the ratio $F(\text{calc})/F(\text{obs})$ in order to minimise the effect of incorrect phases on the convergence.

4) The S-limit for the next cycle to be determined from the R-factors obtained in the two previous cycles, $R(-2)$ and $R(-1)$, from $S=0.2/[R(-2) + R(-1)]$, so that if the refinement started to diverge, the S-limit would be reduced automatically before the next cycle.

5) The volumes of the anisotropic thermal ellipsoids to be constrained.

6) SFLS calculations to be either full matrix or diagonal, although the stereochemical restrictions always involved full matrix calculations.

This modified program proved highly successful in practice, and was described at the I.U.Cr. Congress in 1975 (3).

Experimental Design

The Bayesian approach to the crystal structure analysis problem had, essentially, two separate objectives:

1. To obtain rapid convergence when starting from an approximate trial structure.

2. To decide when convergence has been achieved, and how to assess the result.

It does not really matter how the first objective is reached, but it is important to decide what new chemical information, if any, has been obtained from a particular structure determination.

This requirement suggested a further objective:

3. To treat data collection itself as a design problem, and to identify those reflexions which should be measured or remeasured in order to improve the precision of the chemical data actually being sought.

The specific case considered originally was to decide which additional 100 diffractometer measurements should be made in order (a) to improve the accuracy of the location of the hydrogen atoms in anthracene obtained using the Mo$K\alpha$ diffractometer data measured by D.C. Phillips in 1957 and used in subsequent refinements of the anthracene structure (4,5), or (b) to improve the positional parameters of the 9:10 carbon atom, which in earlier refinements had appeared to show possibly significant deviation from the mean molecular plane, but the design aspect of the work was not pursued further at the time.

It is also possible to utilise this approach in the selection of rotation axes and oscillation ranges to maximise the efficiency of photographic data collection in cases where a small number of reflexions whose identity can be known before data collection commences suffice to establish, for example, site occupancy in solid-solution systems (6).

Both types of example mentioned above presuppose that the information on which the design is to be based is available before the data collection process is initiated, so that no interactive procedure is involved, but now that diffractometers are routinely connected to dedicated computers it is not only possible, but far more cost-effective, to design interactive procedures that can actually modify the data collection process itself as a consequence of analysing the data already collected, so that data collection ceases when the stereochemical objectives have been attained.

Diphenyl Sulphoxide

As an illustration of the progress of a refinement using the specification of the variances of the kind described, we consider the monoclinic crystal of diphenyl sulphoxide (7) having space group $P2_1/n$ with $Z=4$, for which the visually estimated 3-D film data refined to $R=0.174$. A trial structure for this crystal was obtained from diamagnetic anisotropy measurements interpreted on the assumption that the S atom, which had been located from Patterson synthesis, possessed tetrahedral symmetry, so that the angle between the two phenyl groups was expected to be 109° (Fig 1a). In the event it was found to be 97° (Fig 1b), so that the trial structure, although all fractional co-ordinates were within 0.1 of their final values, gave a poor R-factor and incorrect signs for several important reflections, even among the low-angle data.

This is a commonly occurring situation for trial structures derived from stereo-chemical criteria and packing considerations, and was among the first examples shown (2) to be amenable to rapid solution using the procedure described in the Introduction.

The progress of the refinement using the F(obs) values from (7) and the variances shown in Table I is shown in Table II.

The main features of interest in Table II are the large changes in fractional co-

Expert Systems for Data Acquisitions

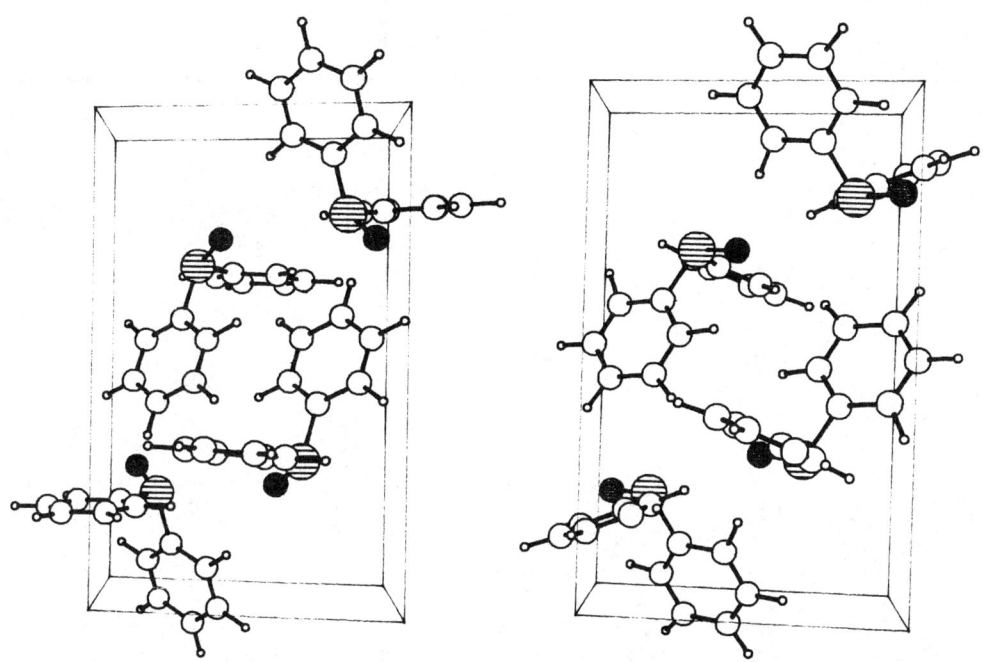

Figure 1a. Trial structure used in the refinement of diphenyl sulphoxide, viewed along c.

Figure 1b. Actual structure obtained from the structure refinement, viewed along c.

ordinates and the range of values taken by the restricted bonds in relation to the variances shown in Table I, and in particular the considerable fluctuations seen in the C—H bonds (hydrogen atoms were not included in the published refinement).

Such fluctuations would not have been permitted by the original procedure (2), where the repositioning of a molecular fragment such as a phenyl group in accordance with the inertia calculations is essentially a type of rigid-group refinement (8).

Table I
Variances and chemical constraints used for diphenyl sulphoxide

Variances for the parameters		Chemical restrictions Å and variances	
Scale factor:	0.05	S—O	1.50 (.001)
S (x,y,z)	0.001	S—C	1.79 (.001)
S,O $(B(ij))$	0.0001	C...C	2.95 (.1)*
O,C,H (x,y,z)	0.05	C...O	2.53 (.1)*
C,H $(B\ iso)$	0.10	C—C	1.40 (.001)
		C—H	1.09 (.001)

*Tetrahedral S requirements
Planarity constraints: .0001

Table II
Progress of the Refinement of the diphenyl sulphoxide

Cycle	S	N	R	R(w)	S—O	S—C	C—C	C—H	$d(x,y,z)$max: S	C,O	H
1	.2000	62	.605	.523	1.61	1.843	1.410	1.14	.0084	.054	.128
						1.848	1.452	2.30			
2	.1810	42	.358	.334	1.55	1.817	1.402	1.10	.0043	.083	.113
						1.827	1.484	1.48			
3	.2077	69	.246	.198	1.497	1.818	1.395	1.11	.0055	.051	.098
						1.811	1.454	1.64			
4	.3311	252	.253	.226	1.474	1.773	1.360	1.11	.0063	.025	.064
						1.756	1.412	1.24			
5	.4006	398	.206	.180	1.478	1.751	1.357	1.089	.0016	.014	.051
						1.767	1.408	1.129			
6	.4352	469	.145	.135	1.489	1.762	1.355	1.065	.0010	.010	.037
						1.780	1.422	1.115			
7	.5688	639	.143	.125	1.490	1.768	1.352	1.055	.0002	.002	.011
						1.784	1.422	1.118			
8	.6940	672	.138	.121							

The other feature of interest is the sudden increase in the value of S, and hence in the number of new reflexions requested for the 4th cycle. It is at this stage that the question of sampling the data in the S-shell instead of collecting them all arises, and will be discussed below.

The first stage of a self-controlled data-collection procedure in this case would have involved the location of the S atom by updating calculations of the appropriate Patterson-Harker lines using successive reflexions (ordered w.r.t. S) until acceptable co-ordinates for the S atom had been established, but the necessary Fourier program has not been yet linked into the SFLS program, and for the trials discussed here the co-ordinates of the S atom were assumed to be known.

The question of how far the detailed analysis of the developing Patterson function could be used during the data-collection process in general has not yet been considered.

Anthracene

Anthracene represents an even more extreme example of experimental overdetermination which cannot be considered cost-effective in terms of diffractometer usage. Since there are only two molecules in a cell of space group $P2_1/a$, the molecule must lie on a centre of symmetry, and may be specified as a planar structure consisting of regular carbon hexagons of edge 1.40Å with associated C—H bonds of 1.09Å, using the same variances as in the case of diphenyl sulphoxide (Table I). Alternatively, if we were to take full advantage of the powerful conformational analysis programs now available, we might specify individual C—C bonds more accurately, and with smaller variances.

Expert Systems for Data Acquisitions

A trial molecule having the observed bond lengths was in fact used for the anthracene trials (Table III), but in the case where the prior information was not altered from the values in Table I, this more detailed information was essentially lost during the substantial changes in fractional co-ordinates required in the first few cycles, although the expected symmetrical variations reappeared later.

Table III
Progress of the Refinement of the Anthracene Structure

Cycle	S	N	R	R(w)	$d(x,y,z)$max: C	H
1	.2000	17	.460	.569	.031	.022
2	.2364	30	.501	.521	.066	.115
3	.1988	16	.100	.095	.017	.032
4	4.962	240	.164	.184	.011	.063
5	1.863	240	.073	.080	.003	.024
6	9.358	240	.064	.067	.0002	.004
7	12.132	240	.061	.064	.00005	.001
8	13.33	240	.061	.064		

The main feature of Table III is the sudden increase in S required in the 4th cycle, after a false start, as in the case of diphenyl sulphoxide (Table II).

The false starts referred to are a consequence of the fact that the original value of S is derived from two estimates of what R-factors would have been obtained in two previous cycles, so that the program has to reach the third cycle before it actually takes control of the number of reflections used. Values of $R(-2)$ and $R(-1)$ of 0.5 give $S = 0.2$ for the first cycle, and have been used in these examples.

Anthracene (20 reflexions only)

Since the molecule is located on a centre of symmetry and may be considered as a rigid planar group, there are essentially only three variables (the orientation angles w.r.t. the crystal axes), and packing considerations indicate that the molecule must lie approximately parallel to the c-axis (information already available from refractive index and diamagnetic susceptibility measurements as well); anisotropic thermal parameters were allowed in order to test the efficiency of the prior restrictions for such low-angle data, so that there were 79 variable parameters in all, and the same prior restrictions on the stereochemistry as before.

A refinement carried out with the first 20 reflexions only ($S = 0.2145$) converged after 8 cycles to $R = 0.009$, but this value is spuriously low, since $R = 0.032$ for the same reflexions in the complete refinement using the available 240 reflexions. Nevertheless the average discrepancies between the two sets of fractional co-ordinates x,y,z (0.010, 0.007, 0.006) are small enough so that a real-time "packing structure" for anthracene could have been obtained in less than 30 minutes of diffractometer time.

Mesogens and Magnetism

An important class of structures for which this procedure is intended are mesogens (liquid-crystal precursors) where the intermolecular forces near the transition temperature are unlikely to be strong enough to affect the geometry of the molecules themselves, except in regard to rotation about single bonds, if any exist.

Existing stereochemical knowledge therefore suffices to provide a trial structure, and the design problem becomes one of collecting sufficient data to establish the position and orientation of the molecule (six parameters only in favourable cases), which in principle can be accomplished using a very small number of reflections.

A diamagnetic susceptibility data base has recently been constructed for use in association with the molecular geometry program RALPuto78 (9), so that diamagnetic anisotropy calculations can be obtained for molecules of any specified geometry. Diamagnetic anisotropy measurements are quick and easy to make, and it is our aim to combine magnetic information with stereochemical knowledge in an expert system to speed up the determination of mesogen structures using controlled data collection procedures of the type to be described.

Data Sampling

The situation shown in Tables II and III where the program indicates that all data within the limiting sphere can now be included in the refinement process raises the question of how much of it would add any useful information to that which we already have. This situation is amenable to statistical analysis, and is discussed by one of us (C.M.O'B.) in Appendix 2.

In order to test the practical efficacy of some simple options of this type, the existing SFLS program has been modified by one of us (M.J.M.) to permit the following operations:

1) Reflexions can be added to the stereochemical information one at a time, ordered w.r.t. S, until further reflections do not shift the molecule appreciably. This option does not solve the problem of increasing the resolution without unreasonably increasing the number of reflexions involved.

2) The data within a particular S-shell can be randomised, and the first half used to refine the structure, which can then be tested against the second half. This option offers information about the quality of the refinement, but again does not limit the quantity of the data collected.

3) The adoption of a procedure for increasing the resolution without measuring all available reflections by specifying: (a) $N(s)$, the number of reflections to be used for the first cycle; (b) $N(i)$, the number of reflections by which the list is to be incremented after each cycle; (c) P, the percentage of reflections in the

Expert Systems for Data Acquisitions

next S-shell which are to be chosen from the randomised list of reflections in this shell; (d) R, the number of cycles to be used with the complete data set when the limit of S has been reached in order to obtain convergence (a decision which will eventually be taken by the program itself).

In this case the program controls the value of S as before, and while, for the rth cycle, $N(r) < [N(s) + (r-1)N(i)]$, all reflections are collected. When $N(r)$ determined by $S(r)$ exceeds this value, the percentage option operates to determine which reflexions are to be collected.

At present these additional reflexions are selected at random, but the current parameters could be used to calculate the expected values of $F(hkl)$ for the unmeasured data in order to select reflections in various ways as required by the procedures envisaged in Appendix 2. It would then be possible to identify, for example, the strongest reflections, since these could be measured most accurately and rapidly; those Friedel pairs showing the largest intensity variations; those strong high-order reflections likely to lead to improved cell parameters; reflexions especially sensitive to the crystallographic information being sought—e.g. site occupancies, hydrogen positions or evidence of static or dynamic disorder.

An option of this general design is well suited to the data-collection process, since $N(i)$ can be chosen in relation to the real times involved on the one hand in re-orienting the crystal and measuring the data, and on the other hand in computing a cycle of refinement and obtaining hard copy or VDU displays of relevant information.

When the standard deviations of the fractional co-ordinates were expressed in Ångstroms it could be seen that to a first approximation $V(x) = V(y) = V(z)$, so that we can obtain bondlength variances $V(ij)$ from $V(ij) = V(i) + V(j)$ (10). Using this approximation, variances of the actual bond lengths after the refinement has converged can be compared with the variances assigned to the prior information (which is a possible criterion which could be used to decide whether additional data should be collected.)

In these anthracene tests an equivalent percentage of the remaining data was randomly selected and refined using the parameters from the converged refinement as the new prior (but without adjusting the variances in accordance with the approximation mentioned above) in order to see how well the refinements using different randomly selected reflexions but the same prior agreed with one another.

Fig. 2a shows results using the rough prior (Table I) of the sort likely to be available for most molecules, and Fig. 2b shows results for tight prior on the bond lengths ($V = 0.00005$) available in favourable cases such as anthracene.

Tests of this procedure on the refinement of anthracene using Phillips' data produced satisfactory convergence after the incorporation of about 80 reflexions

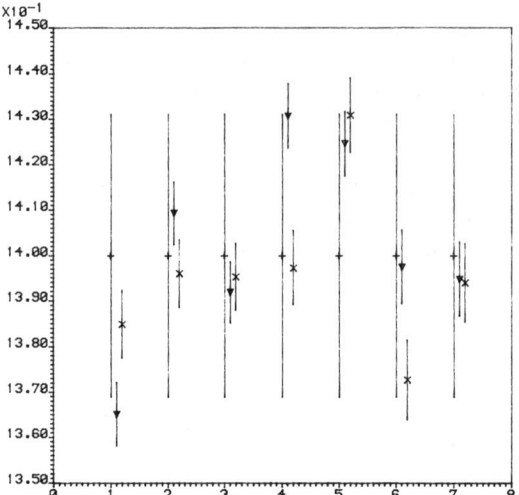 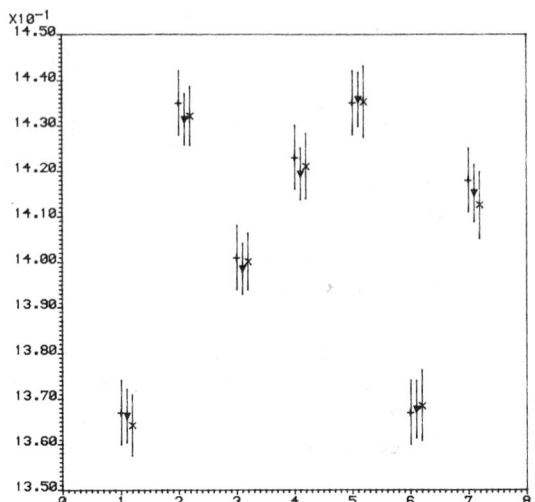

Figure 2a. Values of bond-lengths in anthracene obtained from refinements using partial data sets, with bond lengths and variances as specified in Table I. — = Table I, × = diagonal SFLS, ∇ = full matrix SFLS.

Figure 2b. Values of bond-lengths in anthracene obtained from refinements using partial data sets and more accurate bond-lengths and variances. — = Sparks (4), × = diagonal SFLS, ∇ = full matrix SFLS.

(correponding to about 1-2 hours of diffractometer time), but this is not in fact a valid test, because Phillips' data exclude some of the weaker reflexions available out to the S-limit, and thus already incorporate one of the possible refinement criteria mentioned above. The examples given are intended merely to suggest a new approach to the problems of devising expert systems to control the acquisition of crystallographic data in the light of available stereochemical information and defined stereochemical objectives. Detailed development of these ideas will be published elsewhere.

Acknowledgement

One of us (C.M.O'B.) acknowledges the financial support of the Science and Engineering Research Council.

Appendix 1: analytic formalism involved in our approach to crystal structure determination

The basic theory pertaining to particle beam diffraction by crystals is well-known and familiar to crystallographers—texts such as (11,12) providing a comprehensive treatment of the topic of crystal-structure analysis both for the novice and experienced research worker.

In brief, the essential premise of a crystal-structure determination is that the *unknown* structure of a crystal may be characterized by diffraction spectra hkl. Denote a set of diffraction spectra by $\{h_j k_j l_j; j=1,...,m \ (\leq N)\}$ where N is as defined in the *Introduction* to this paper. Crystal-structure analysis consists in determining a model structure whose theoretical diffraction spectra mimic, in a suitably prescribed manner, an observed set of spectra. More specifically, denoting the number of atoms in the unit cell of a crystal specimen (under investigation) by R and the number of measured reflexions by m, the coordinates $\{(x_r, y_r, z_r); r=1,...,R \ (\ll N)\}$ of atoms in a model structure are determined such that the set of *calculated* structure factors $\{F_c(h_j k_j l_j); j=1,...,m\}$—with $F_c(h_j k_j l_j)$ denoting the calculated structure factor for the $h_j k_j l_j$ spectrum—'best fits' an *observed* set of structure factors $\{F_o(h_j k_j l_j); j=1,...,m\}$—with $F_o(h_j k_j l_j)$ denoting the observed structure factor for the $h_j k_j l_j$ spectrum. Note that while much has been written concerning the present-day analyses of crystals—particularly with reference to the phrase 'best fits'—this is neither the time nor the place to review such material.

This *Appendix 1* will address primarily the problem of the incorporation of amassed chemical information into an *estimated determination* of the coordinates $\{(x_r, y_r, z_r); r=1,...,R\}$ of atoms in a crystal of unknown structure. Merely by way of an example, these (coordinates) are typically required to be *consistent* (e.g. in the l_2-norm sense) with the observed amplitudes $|F_o(h_j k_j l_j)|$, for $j=1,...,m$, of measured diffraction spectra.

In general, the structure factor for a generic *hkl* spectrum is computed by a relation

$$F_c(hkl) = \sum_r s_r \exp\{i2\pi(hx_r + ky_r + lz_r)\}, \qquad (A.1)$$

where s_r denotes the *r*th atomic scattering factor, and the *r*th parameter triple (x_r, y_r, z_r) is unknown but to be determined. Such (highly) non-linear functions are at the heart of crystal-structure analysis (13). Estimation of the coordinates $\{(x_r, y_r, z_r); r=1,...,R\}$ of atoms is frequently achieved through a linearization of the equation (A.1) using the first two terms in a Taylor series (11) followed by a least-squares procedure (14). The adoption of such an approach does not permit either the incorporation of previously amassed chemical information or [excepting a few analytic procedures (15)] *prior* parameter information. This deficiency may be corrected by appealing to the *Bayesian* inferential paradigm (16) and presented next.

Suppose we have a known *response* function analogue of the equation (A.1)

$$\eta = f(hkl; \theta_1, \theta_2, ..., \theta_p)$$
$$= f(hkl, \theta), \qquad (A.2)$$

where the $1 \times p$ vector $\theta^T = (\theta_1, \theta_2, ..., \theta_p)$ represents values of p parameters to be estimated; e.g. fractional positional coordinates of all the atoms (usually three per atom), isotropic or anisotropic temperature factor parameters (up to six per atom), an overall scale factor, and possibly fractional occupation numbers if some atoms are *stochastically* distributed over several sites. Suppose we have *m observations*

$$y_j = \eta_j + \epsilon_j$$
$$= f(h_j k_j l_j, \theta) + \epsilon_j \quad (j=1,\ldots,m), \tag{A.3}$$

where the *disturbances* ϵ_j are independently normally distributed with zero mean and variance σ^2—an alternative description of the supposition (A.3) being

$$y_j \sim N[f(h_j k_j l_j, \theta), \sigma^2]. \tag{A.4}$$

Assume that, for a region in the θ space sufficiently close to any selected value of θ, say θ^*, the following expansion is valid

$$f(h_j k_j l_j, \theta) \simeq f(h_j k_j l_j, \theta^*) + \sum_{t=1}^{p} (\theta_t - \theta_t^*) g_j^{(t)}, \tag{A.5}$$

$$\text{where } g_j^{(t)} = \left. \frac{\partial f(h_j k_j l_j, \theta)}{\partial \theta_t} \right|_{\theta=\theta^*}. \tag{A.6}$$

Substitution of the expansion (A.5) into the equation (A.3) leading to the formation of an *approximate* vector equation

$$a = G(\theta - \theta^*) + u, \tag{A.7}$$

$$\text{where } a = \begin{bmatrix} a_1 \\ a_2 \\ \vdots \\ a_m \end{bmatrix}, \; G = \begin{bmatrix} g_1^T \\ g_2^T \\ \vdots \\ g_m^T \end{bmatrix}, \; u = \begin{bmatrix} u_1 \\ u_2 \\ \vdots \\ u_m \end{bmatrix},$$

$$\left. \begin{array}{l} a_j = y_j - f(h_j k_j l_j, \theta^*) \\ \\ g_j^T = (g_j^{(1)}, \ldots, g_j^{(p)}) \end{array} \right\} \; ; \text{ for } j=1,\ldots,m,$$

and with each u_j representing the amalgamation of the Taylor series residual of higher order (2nd, 3rd,...) terms and disturbance ϵ_j. Assume without loss of generality that

$$u \sim N_m(O, C), \tag{A.8}$$

with $N_m(O, C)$ denoting the m-dimensional multivariate normal distribution with mean vector O and dispersion matrix C (assumed *known*). Defining

$$\phi = (\theta - \theta^*) \tag{A.9}$$

and using (A.8) in conjunction with (A.7) leads to the distributional result that, given ϕ,

Expert Systems for Data Acquisitions

$$a \sim N_m(G\phi, C). \tag{A.10}$$

Suppose that as an investigator *my prior beliefs* about the parameter vector θ (and by implication ϕ) before obtaining m observations $y_1, y_2, ... y_m$ may be conceptualized in the following manner. Given ϕ_1,

$$\phi \sim N_p(A\phi_1, C_1) \tag{A.11}$$

and given ϕ_2,

$$\phi_1 \sim N_q(B\phi_2, C_2). \tag{A.12}$$

The dispersion matrices C, C_1 and C_2 have obvious dimensions and are supposed non-singular. Then, as shown in (17), the posterior distribution of ϕ, given $G, A, B, C, C_1, C_2, \phi_2$ and $y^T = (y_1, y_2, ..., y_m)$ is $N_p(Dd, D)$ with

$$D^{-1} = G^T C^{-1} G + \{C_1 + AC_2 A^T\}^{-1} \tag{A.13}$$

and

$$d = G^T C^{-1} a + \{C_1 + AC_2 A^T\}^{-1} AB\phi_2. \tag{A.14}$$

Trivial algebraic manipulation then giving the *posterior* mean—denote $\theta^{(1)}$—and dispersion matrix for the parameter vector θ, conditional on the value of θ^*. The approximate linearization (A.5) may be repeated with $\theta^{(1)}$ in place in θ^* and a new posterior mean, $\theta^{(2)}$, and dispersion matrix (both conditional on the value of $\theta^{(1)}$) calculated. The process of substitution and re-estimation may be continued *ad infinitum* or until such time as the difference between successive (iterative) estimates of the vector θ become *small* [in an arbitrary but specified sense (18)]. The mean of the posterior distribution obtained when the iteration procedure converges is taken as one's *estimate* of the parameter vector θ.

Appendix 2: proposed development of our approach

There follow outline statements of some open problems together with citations of pertinent published works. We hope to develop solutions to these problems and also to propose a coherent approach to the problem of experimental design and data analysis for crystal structure determination.

1. *Can a parametrization of the model (A.4) be found for which inference about a subset of the parameters $\theta_1, ..., \theta_p$ is simple?* For example, one may be interested in the fractional positional coordinates of all the atoms in a model structure but not the isotropic (or anisotropic) temperature factor parameters—which one may regard as *nuisance* parameters. Recent work (19) alluding to the existence or otherwise of a suitable transformation of a model (A.4) may be adapted to obtain an answer to this question.

2. *Can the Bayesian procedure be modified to incorporate non-linear relationships in the specifications (A.11) and (A.12)?* The replacement of the linear components in the parameters $\theta_1,...,\theta_p$ in the mean vectors of (A.11) and (A.12) by non-linear functions may be accommodated within the proposed Bayesian procedure by the use of Taylor series approximations (*c.f.* the specification (A.5)) *and* the appropriate augmentation of the iteration routine described in the *Appendix 1* to this paper. By way of an example to illustrate how a non-linear parameter specification may arise, suppose that one is interested in the bond length between two atoms—a non-linear function of the six atomic coordinates. Then, for the case of a group of similar carbon-carbon bonds, the ϕ of (A.11) may be the vector of carbon-carbon bond lengths while the ϕ_1 of (A.12) may be the common mean.

3. *Given that one may order the N available reflexions and partition them into a finite number of non-overlapping intervals (or strata) on the basis of $\sin\theta/\lambda$, how many reflexions need one measure within each stratum?* Appealing to the sampling techniques of (20) and the developments in (21) concerning data selection may provide a solution to this problem. Developments in (22) concerning methods that intentionally ignore some of the accessible data may also prove helpful.

4. *Given measurements on m reflexions, how should additional reflexions be selected in order to obtain the 'best' estimation of the parameters $\theta_1,...,\theta_p$?* In (23) one sense of the term 'best' is proposed and investigated with the use of prior distributions in the design of experiments for parameter estimation in non-linear situations. The extension of these ideas into the area of crystal structure analysis appears to be possible. However, attention to the problem of data-dependent stopping rules raised in (24) may be necessary.

5. *Given various heterogeneous data sets (each containing a mean and standard deviation for a number of crystallographic parameters), how may one incorporate all these opinions into a single, reconciled probability distribution for (i) each parameter, and (ii) all parameters jointly?* Recent developments in (25) go some of the way to providing an answer to this question but further research (with particular emphasis on the problems encountered in crystal structure determination) is still required.

6. *Can one design a data collection procedure which does not measure all of the available N reflexions but minimizes the cost of collecting the data while achieving desired precision of parameter estimates?* Results in (26) suggest that one can, and offer the distinct possibility of being able to develop a practicable design for use in crystal structure determinations if different costs can be assigned to different data points.

In conclusion, the Bayesian procedure described in this paper represents merely one of the many powerful and useful tools at the statisticians' disposal with which to achieve the objective of incorporating prior parameter information into a crystal structure determination (with a view to influencing the data collection process). This suggests that great gains may be obtainable by attention to the *design* and *controlled* acquisition of experimental data.

References and Footnotes

1. G.I. Webb, unpublished paper presented at a one day symposium on *Bayesian Methods in Crystallography* held in Oxford (1974). See also S. French, *Acta Cryst. A34,* 728 (1978).
2. H.J. Milledge, *Proc. Roy. Soc. A267,* 566 (1962).
3. M.J. Mendelssohn, H.J. Milledge and G.I. Webb, *Acta Cryst. A31,* S232 (1975).
4. R.A. Sparks, PhD Thesis, University of California, Los Angeles (1958).
5. D.W.J. Cruickshank and R.A. Sparks, *Proc. Roy. Soc. A258,* 270 (1960).
6. H.J, Milledge, *Acta Cryst. A25,* 173 (1969).
7. S.C. Abrahams, *Acta Cryst. 10,* 417 (1957).
8. C. Scheringer, *Acta Cryst. 19,* 513 (1965).
9. G. M. Crisp, Rutherford and Appleton Laboratory Central Computing Division: RAL Pluto 78 Users' Guide (1984).
10. H. Lipson and W. Cochran, *The Determination of Crystal Structures.* G. Bell & Sons Ltd., London, p. 309 (1953).
11. M.J. Buerger, *Crystal-Structure Analysis.* New York, Wiley (1960).
12. J. D. Dunitz, *X-Ray Analysis and the Structure of Organic Molecules.* Ithaca, New York, Cornell University Press (1979).
13. J. S. Rollett, *Computing Methods in Crystallography.* Oxford, Pergamon Press (1965).
14. J.R. Wolberg and J. Isenberg, *Nuclear Instruments and Methods in Physics Research 112,* 533 (1973).
15. H. Toutenburg, *Prior Information in Linear Models.* Chichester, Wiley (1982).
16. D.V. Lindley, *Bayesian Statistics, A Review.* Society for Industry and Applied Mathematics, Philadelphia, Pennsylvania (1971).
17. D.V. Lindley and A.F.M. Smith, *Journal of the Royal Statistical Society, Series B, 34,* 1 (1972).
18. J.M. Chambers, *Biometrika, 60,* 1 (1973).
19. P. Hougaard, *Parameter Transformations in Multiparameter Nonlinear Regression Models.* Preprint no. 2, Institute of Mathematical Statistics, University of Copenhagen (1984).
20. W.G. Cochran, *Sampling Techniques;* 3rd edition. New York, Wiley (1977).
21. W.A. Ericson, *Journal of the American Statistical Association 60,* 750 (1965).
22. D.B. Rubin, *Biometrika 63,* 581 (1976).
23. N.R. Draper and W.G. Hunter, *Biometrika 54,* 147 (1967).
24. P.R. Rosenbaum and D.B. Rubin, *The American Statistician 38,* 106 (1984).
25. D. Lindley, *Operations Research 31,* 866 (1983).
26. R.R. Hocking and W.B. Smith, *Technometrics 14,* 299 (1972).

INDEX

Abrahams, S.C., 135, 195, 211
   sources of error, 126, 127, 128-129
   structure of tartaric acid, 183, 187-190
Absolute scale
   derivation from relative, 50
Abramowitz, M., 42, 66, 94
   Bessel functions, 29
Accuracy, contrasted with precision, v, 95-107
Acentric distribution
   Fourier representation, 26-32
   orthogonal polynomials and, 90
Acetyl histidine monohydrate, 85
Ahmed, F.R., editor, 22
Ahmed, R., vi, 181
   author, 151-181
   data used, 174
Albinati, A., 103
Alexander, L.E., 181
Ammon, H.L., 136
Anisotropy, diamagnetic, 200
Anomalous-dispersion method, contrasted with direct, 2
Anorthite, 48, 50
Anthracene, 200, 202-203
Antimony lead silver sulphide, 74
Asymptotic approximations
   of intensity distributions, 32-34, 88
Atomic heterogeneity, *see* Heterogeneity
Atomic scattering factors
   and absolute scale, 3
   computed by Cromer and Mann, 188
   computed by Forsyth and Wells, 188
Avrami, M., 21
   relation of intensities to interatomic vectors, 2

Background
   effect on intensity distributions, 88, 90
   refinement of level of, 174
Baharie, E., 103
Banerjee, K., 21
   relations between structure factors, 2
Barakat, R., 42
   Fourier-series representation of distributions, 25
Barakat series, 25, 87
Barton, R.J., 136
Bassi, G.S., editor, 136
Bayes' theorem, 192
Bayesian
   approach to structure determination, 198-200, 210
   interpretation of unobserved reflections, 193
Beaumont, G.P., 85
Berg, P.W., 42
   Fourier-Bessel series, 29
Bertaut, E.F., 42
   identity used in direct methods, 39
Bessel functions
   modified, 33, 45, 88-89
   of first kind, 29, 60
   zeroes of, 29
Best tests, 79-85
Beu, K.E., vii, 181
   likelihood-ratio test, 166
   maximum-likelihood method, 153, 154-155
   normal distribution of angular errors, 157
Bias
   in parameter estimation, v, vi, 95, 107
   See also Systematic error
Biocentric distribution
   [special case of hypercentric], 6
   Fourier representation of, 53-66 (especially 59-60)
   Hermite polynomial representation of, 54-55
Bloss, F.D., 51
Boehme, R., 77

recognition of pseudo-
   translations, 73
Bond, W.L., 137
   method, 137-149
Bond density, inadequacy of
   model for, 126
Bond lengths
   anthracene, 206
   diphenyl sulphoxide, 202
Bond method for lattice
   parameters, vi, 137-149
Box, G.E.P., 195
Bricogne, G., 123
   formulation of direct methods,
      118
Brill, R., 21
3-Bromo-1,8-dimethylnaphthalene,
   47-49
Bryan, R., 124
Buchanan, J., 136, 195
Buerger, M., 77, 211
   classification of superstructures,
      67, 72
Bullock, J.I., 66

Caglioti, G., 103
   peak parameters, 103
Cascarano, G., v, 77
   author, 67-77
Central limit theorem
   limitations of, 23
   probability distributions derived
      from, 14, 54, 76, 90-91, 138
Centric distribution
   Fourier representation of, 25-26,
      53-66
   orthogonal polynomials and, 90
Centrosymmetric distribution, *See*
   Centric distribution
Centrosymmetry
   partial, 36-37
   statistical tests for [absence of],
      1, 5, 79-84
Chambers, J.M., 211
Characteristic functions
   in probability-distribution

   theory, 23-41, 53-66
   for specific space groups, 34-37
Chaudhuri, A.K., 181
   moment-range plots, 177
Chebyshev functions, method, 156
Cheng, B-K, 123
Chi-squared, method of minimum,
   151
Christ, C.L., 21
   colemanite, 4
Closed forms
   for intensity distributions, 87,
      89-90
Cocaine, 131-133
Cochran, W., 21, 42, 211
   approximate formula of, 40-41
   multivariate distributions, 37
   phase determination, 12, 14
   probability of tangent formula,
      18, 19
Cochran, W.G., 211
Cole, J.E., 42
   Fourier representation of
      distributions, 25
Colemanite
   illustrating statistical methods, 4
Collin, R.L., 50
   heavy-atom effect, 43
Collins, D.M., vi, 112
   author, 109-112
Convolution, 91
Cooper, M.J., 103
   two-stage Rietveld method, 99
Coppens, P., 135
   inadequacy of atom model, 126
Copper, 158-161
Correlation, method of maximum,
   151
Correlation between atomic
   positions
   effect on distributions, v, vi, 44
Counting statistics, *see* Poisson
   distribution
Covariance [*not exhaustively
   indexed*]
   in Bond method of parameter

Index

determination, 137-149
Cox, G.W., 195
   tartaric acid, 188
Cramér, H., 94, 181
   on approach to normality, 155
Crisp, G.M., 211
Critical region, 80-84
Cromer, D.T., 195
   atomic scattering factors, 188
Cross-entropy, 'relative information' preferred term, 114
Cruickshank, D.W.J., 107, 136, 211
   statistics of residuals, 106
   weighting scheme, 127
Crystallographic computing, in direct methods, 20
Crystallographic statistics
   and direct methods, 1-22, 37-41
   development of, 3-6
   Fourier representations, 23-41
   history of, 1-22
Cumulants, 94
Cumulative distribution function(s) [*not exhaustively indexed*]
   derivation of, 5, 44
   effect of heavy atom(s) on, 43-51
   in 'best tests', 83
   pseudotranslations and, 76-77
Cyclophane triene, 85

Dam, J., 135
Daniell, G.J., 123
   weak constraints, 120
Daniels, G.E., 42
   steepest descents, 26
Das Gupta, P., vi
   author, 151-181
Data acquisition, optimization of, 194, 197-211
Data base, diamagnetism, 204
Data sampling, 204-206
Design matrix, partitioning of, 100
Dickinson, C.W., 136
Dimethyl ester of mesotartaric acid, 85
Direct methods
   general formulation of, 118
   history of, 1-22
   joint distributions in, 37-41
   pseudotranslations and, 77
Directed divergence, 'relative information' preferred term, 114
Debye-Waller factor, *see* Temperature factor
Decaborane, 9
Defects in the model, vi
Determinantal inequalities
   geometrical interpretation, 11
   in structure determination, 8-12
   probabilistic theory, 14-16
Diamagnetism
   anthracene, 203
   data base, 204
   diphenyl sulphoxide, 200
   mesogens, 204
Diphenyl sulphoxide, 200-202
Disorder, positional, 1, 3
Distributions, *see* Intensity distributions
Draper, N.R., 211
Dunitz, J.D., 211

Edgeworth series
   contrasted with Fourier, v, 26, 87-93
   cumulant form of, 94
   in curve fitting, 155
   multivariate, 37
Elango, N., author, 79-85
Elderton, W.P., 181
   criterion for curve fitting, 165
Electron density
   in superstructures, 67
   non-negativity and direct methods, 1
Entropy maximization, 109-112
'Epsilon' factors, 4
Ericson, W.A., 211
Errors
   in angular scale, 138 *et seq.*
   in temperature, 126, 141-142
   in timing, 141

in scanning mechanism, 126, 138 *et seq.*
See also Bias, Random error, Systematic Error
Excluded volume(s), 43-51
Expert systems, 197-211
Extinction, as a systematic error, 87

Faggiani, R., 66
False minima, 192
Fedorov, V.V., 195
    theory of optimal experiments, 186
Feller, W., 42
    Tauberian theorem, 33
Finger, L.W., 195
Fisher, R.A., 181
    maximum-likelihood method, 153
Flack, H.D., 136
    effect of serial correlation, 127
Fluctuations
    in minimum-information methods, 118-119
    statistical, *see* Statistical fluctuations
Forsyth, J.B., 195
    atomic scattering factors, 188
Foster, F., 66
Fourier series
    representing probability distributions, v, 23-32, 37-41, 87, 93
    unpromising for representing fluctuations, 93
Fourier-Bessel series
    representing probability distributions, 29, 53-66, 87-93
French, S., 50, 195
    correlation between atomic positions, 44
    treatment of weak reflections, 191
Friedel pairs, 205
Frieslebenite, 74-77G

Gabe, E.J., 136

Gałdecka, E., vi, [as Urbanowicz] 149
    author, 137-149
Gamma function, 90
Gauss-Markov theorem, 96-97, 100
Gaussian distribution [*not exhaustively indexed*]
    central-limit approximation, 23
    statistical fluctuations approximated by, 89, 127
Generalized intensity statistics, 23-41, 53-66
    See Intensity statistics
Geometric distribution, 90
Ghosh, S., v, 66, 181
    author, 43-51
    thermal parameters for hydrogen, 177
Ghosh, T.B., 181
    data used, 177
Giacovazzo, C., v, 42, 77
    author, 67-77
    joint distribution functions, 37
Gillis, J., 21
    inequalities, 9
Goedkoop, J.A., 21
    determinantal theory, 15
Good, I.J., 123
Goodness-of-fit, defined, 127
Gradshteyn, I.S., 42, 66
    series expansions, 38
Gram-Charlier series
    contrasted with Fourier, v, 26
    in curve fitting, 155
    multivariate, 37
    See also Edgeworth series
Group theory, 'epsilon' factors, 4
Gull, S.F., 123
    weak constraints, 120
Gupta, P.D., vi
    author, 151-181

Hall, S.R., editor, 22
Hamilton, W.C., 195, 135, 136
    bias in parameters, 128
    editor, 136

Index

structure of tartaric acid, 183, 187-190
Hanson, A.W., 85
Hargreaves, A., 21, 66
   effect of heavy atoms, 6
Harkema, S., 135
Harker, D., 21
   decaborane, 9
   inequalities, 9
Hartley, H.O., 181
   criterion for curve fitting, 165
Hauptman, H., 21, 41, 42, 66, 124
   determinantal inequalities, 9, 15-16
   'epsilon' factors, 4
   higher-order terms in distributions, 6
   multivariate distributions, 37
   overdeterminancy, 8
   phase determination, 6, 13, 14
   probability theory, 13
   random walk, 21, 23
   series expansions, 23
   statistical tests for centro-symmetry, 5
Heaviside step-function, 30
Heavy-atom method, contrasted with direct, 2
Heavy atoms, effect on intensity statistics, v, 1, 43-51 *passim,* 53-66
Heck, H., 136
Heinerman, J.J.L., 21
   determinantal probability distributions, 19
Henry, N.F.M., editor, 42
Hermite polynomials, in probability distributions, 54, 91, 94
Heterogeneity, atomic, effect on intensity distributions, 90-91, 93
Hewat, A.W., 103
   absorption and multiple scattering, 103
Hirshfeld, F.L., 195
History
   of crystallographic statistics and direct methods, 1-22

Hoaglin, D.C., 195
Hobson, A., 123
Hoel, P.G., 85
Hong, W., vi
   author, 125-136
Hougard, P., 211
Howells, E.R., 21
   statistical tests for centro-symmetry, 5
Hyrnchuk, R.J., 136
Huber, C.P., editor, 22
Huml, K., editor, 22
van Hummel, C.J., 135
Hunt, D.J., 85
Hunter, W.G., 211
Hydroxyproline, 12
Hyperbolic tangent formula, in phase determination, 14, 19
Hypercentric distributions, 6
   orthogonal polynomials and, 90

Ibers, J.A., editor, 136
Inaccessible volume(s), effect on intensity statistics, 43-51
Inequality theory
   determinantal inequalities, 8-13
   direct methods, 1, 2
   Harker-Kasper, 9, 12
   non-negativity and, 8-13
   probabilistic interpretation, 14-19
Information capacity, as entropy, 109
Information theory, in structure determination, 109-112, 113-124
Integrated intensity, in two-stage Rietveld method, 99-101
Intensity distributions, tests to distinguish, 47-50, 79-85
Intensity statistics
   effect of heavy atom, 1, 23-41 *passim,* 43-51
   generalized, 23-41, 53-66
   history of, 1-22
International Tables for Crystallography, 4, 21

Introduction, v-vii
Invariants
  phase invariants, 19
  seminvariants, 19
Isenberg, J., 211
Isomorphous-replacement
  method, contrasted with direct, 2
Ito, T., 21, 77
  'epsilon' factors, 4
Iwasaki, H., 21
  'epsilon' factors, 4

Jameson, M.B., 51
Jaynes, E.T., 112, 123
  and entropy, 109
  minimum-information principle, 114
Jeffery, J.W., 77
Johnson, R.W., 112, 123
  argument for minimum-information principle, 115
Joint probability distributions
  and direct methods, 1, 13-14
  Fourier representation of, 37-41

Kaldor, U., 42, 66
  moments of structure factor, 24
Kanters, J.A., 85
Karle, I.L., 21
  probability of tangent formula, 18
  tangent formula, 14
Karle, J., v, 21, 22, 41, 42, 66
  author, 1-22
  determinantal inequalities, 9, 15-16
  discussion of structure problem, 7
  'epsilon' factors, 4
  higher-order terms in distributions, 6
  multivariate distributions, 37
  overdeterminancy, 8
  phase determination, 6, 14
  probability of tangent formula, 18
  random walk, 21, 23
  series expansions, 23
  statistical tests for centrosymmetry, 5
  tangent formula, 14
Kasper, J.S., 21
  decaborane, 9
  inequalities, 9
Kay, M.I., 195
  tartaric acid, 188
Kempster, C.J.E., 51
Kendall, M.G. [Sir Maurice], 66
Kiefer, J.E., vii, 42, 66, 94
  accelerated convergence, 26
  author, 23-42
  convergence, 29
  distribution of large structure factors, 32
  steepest descents, 26
Kistenmacher, J.J., 85
Kitaigorodskii, A.I., 42
  partial centrosymmetry, 36
Klimko, E.M., translator, 195
Klug, A., 42, 66
  multivariate distributions, 37, 41
  orthogonal-polynomial expansion, 23
Klug, H.P., 181
Kluyver, J.C., 42
  probability distributions, 24
van Koningsveld, H., editor, 136
Krabbendam, H., 21
Kroon, J., 21, 85
Kucharaczyk, D., 149
Kullbach, S., 123
  information theory, 113, 114

Lack of fit in Rietveld method, 98-99
β-Lactam, 151, 170-177
Ladd, M.F.C., 66
  editor, 124
Lagrange multipliers, 122-123
Laguerre polynomials, 91, 94
Large intensities, v, 32-34
Larson, A.C., 136

Index

Laplace transforms, 33
Laser speckle, analogy with random walk, 25
Lattice constant (parameter)
   determination by Bond method, 137-149
   determination by maximum likelihood, 151-181
   determination by min-max method, 151-181
   distribution of, 159-160
   of copper, 169-170
   of HgClBr, 151, 170-174
   of β-lactam, 151, 170-174
Lead, 151, 177-180
Least squares
   contrasted with entropy maximization, 109
   contrasted with maximum likelihood, 153-155
   contrasted with min-max method, 155-156
   incorporation of chemical information, 197
   modification of weights, 125-135
   refinement of parameters, 95, 152, 193-195
Lehmann, M.S., 123
Leverage
   of data points, defined, 185
   of weak reflections, 187
Levine, R.D., 112, 123
Lide, D.R., editor, 195
Likelihood, 80, 192-193
Lindley, D.V., 211
Lippert, B., 66
Lipson, H., 21, 42, 66, 77, 94, 211
   hypercentric [bicentric] distribution, 6, 24, 59
Liquid-crystal precursors, 204
Lock, C.J.L., 66
Lonsdale, K., editor, 42
Lorentz factor, 191, 194
Lucht, C.M., 21
   decaborane, 9
Luić, M., v, 77
   author, 67-77
Łukaszewicz, K., 149
McGregor, J.L., 42
   Fourier-Bessel series, 29
Machin, P.A., 136
Mackenzie, J.K., 195
   extinction correction, 183, 187
McLachlan, A., 124
Magnetism, 204
Malinowski, M., 149
Mann, J.B., 195
Marsh, R.H., 85
Mathieson, A.McL., 195
Maximum-entropy principle, equivalent to minimum information, 113
Maximum-likelihood method, 151-181
   contrasted with least squares, 153-155
Megaw, H.D., 51
Mendelssohn, M.J., vi, 211
   author, 197-211
Mercury bromide chloride, 151, 170-177
Mesogens, 204
Metaboric acid, 13
Meyer, P.L., 85
Milledge, H.J., vi, 211
   author, 197-211
Minimum-information principle, equivalent to maximum entropy, 113
Min-max method, 151-181
   combined with maximum likelihood, 156-157
   contrasted with least squares, 155-156
   in structure refinement, 180
Misra, N.K., 181
   moment-range plots, 153, 177
   particle size and strain, 177
   refinement of background level, 174-177
Mitra, G.B., vi, 181
   author, 151-181

moment-range plots, 153, 177
particle size and strain, 177
refinement of background level, 174-177
thermal parameters for hydrogen, 177
Model, mathematical, for superstructures, 68-72
Modified variance, 129-130
Moments, method of, 151
Moments of line profile
in determining crystallite size and strain, 151, 177-178
Moments of structure factors
in series coefficients, 91-93, 94
integrals for, 4, 30-31
Monte Carlo method, 82, 83
Mood, A.M., 181
maximum likelihood *versus* least squares, 155
Mukherjee, P.S., 181
moment-range plots, 153, 177
Musil, F.J., vii, 181
likelihood-ratio test, 166
maximum-likelihood method, 153

Narrow-band noise, analogy with random walk, 25
Navaza, J., 124
Nelson, J.B., 181
extrapolation technique, 153, 169
Neyman-Pearson lemma, 79-82
Nicholson, W.L., vi, 136, 195
author, 183-195
Nielson, K., 136
weighting scheme, 127
Nigam, G.D., v, 50, 51, 66, 181
author, 43-51
thermal parameters of hydrogen, 177
Noncentrosymmetric distribution, *see* Acentric distribution
Non-centrosymmetry, *see* Centrosymmetry

Non-crystallographic symmetry
effect on intensity distributions, 6, 24, 67
recognition of, 72
Normal (Gaussian) distribution
[*not exhaustively indexed*]
central-limit approximation, 23
in maximum-likelihood method, 154-155
statistical fluctuations approximated by, 89, 127
Normalized structure factor, in direct methods, 4
Nowacki, W., 77

O'Brien, C.M., vi
author, 197-211
Observations, meaning of independence of, 101
Okaya, Y., 195
tartaric acid, 188
Olthof-Hazekamp, R., editor, 136
Optimization
of data collection, 184
of experiments, 194
Ott, H., 21
relation of structure factors and atomic positions, 2
Orthogonal polynomials
expansion of series of, 87, 90-93, 94
Hermite, 54, 57, 94
Laguerre, 56, 57, 94
Outlier, defined, 185
Overdeterminancy in structure determination, 1, 3, 7-8, 20, 192, 198
Overlapping peaks
in structure refinement, 95

Palmer, R.A., editor, 124
Paoletti, A., 103
peak parameters, 102
Parameter estimation
by entropy maximization, vi, 109-112

Index

by min-max methods, 151, 180
precision of, v-vi, 183-195
precision and accuracy in, v, 95-107
Parthasarathi, V., 42, 85
partial centrosymmetry, 37
Parthasarathy, S., v, 42, 66, 94
author, 79-85
partial centrosymmetry, 37
Partial centrosymmetry, 36-37
Particle-size determination, 177-180
Partially bicentric distribution, 53, 60-66
Patterson, A.L., 21
function, 7
Patterson function
in solution of structure problem, 7
structure of diphenyl sulphoxide, 200
Patterson-Harker lines, 202
Patterson methods
contrasted with direct, 2
inaccessible volume, 44
Paul, M.A., editor, 195
Pawley, G.S., 103
Pearson, E.S., 181
Pearson, K., 41
random walk, 23
Pearson curves, 155, 161, 163, 178
Penfold, B.R., 51
Pepinsky, R., 42
approximate formula of, 40-41
editor, 136
multivariate distributions, 37
Phase determination
and joint probability distributions, 1
effects of randomness and rational dependence, 6
in minimum-information methods, 120-121
Phillips, D.C., 21
data for anthracene, 200, 205
statistical tests for centro-

symmetry, 5
Pietraszko, A., 149
Piro, O., 123
Pixels, 116
Poisson distribution
combination with ideal distributions, 89-90
in maximum-likelihood method, 155
statistical fluctuations represented by, 88, 127, 137, 138
Polarization factor, 191, 194
Polymer configuration, analogy with random walk, 25
Ponnuswamy, M.N., 85
Port, S.C., 85
Positional disorder, 1
Povey, D.C., 66
Powder methods, parameter estimation by, 95, 151-181
Pradhan, D., v
author, 43-51
Precision
contrasted with accuracy, v, 95-107
in least-squares refinement, 183-195
in maximum-likelihood and min-max methods, 151-181
Price, P.F., 181
Poisson distribution in maximum likelihood, 155
Prince, E., v, vi, 136, 195
author, 95-103, 107, 183-195
Prior distribution
in entropy maximization, 109-112
choice of, 116-118, 121-122
Probability distributions
relation between $p(I)$ and $p(F)$, 88
representations of, 87-94
Probability theory
and direct methods, 1
deviations from ideal distri-

butions, 5-6
Fourier representation of distributions, 23-41
Pseudotranslations
in superstructures, 68 *et seq.*

Rabinovich, D., 195
Radoslovich, E.W., 51
Ralston, A., 181
 min-max method, 156
Ramaseshan, S., editor, vii, 66, 136, 181
Random error(s)
 in general, 152
 in structure determination, 126 *et seq.*
 statistical fluctuations as example, 87
Random walk, analogy with structure factor, 13, 23-24, 54
Randomness, departures from, 6
Rational dependence, in phase determination, 6
Refinement
 by maximum likelihood, 151, 180
 modification of weights in, vi, 125-135
 of parameters by least squares, 183-195
 Rietveld method, 95-107
Reflections
 influence of individual on parameters, 183-195
Refraction correction, 170, 171, 172, 173, 175
Relative entropy, 'relative information' preferred term, 114
Relative information, measure for, 114-115
Residual error, 125-135
Residuals
 in Rietveld method, 101
 statistics of, 106
Reuvers, A.J., 135
Ricci, F.P., 103

peak parameters, 102
Richardson, M.F., editor, vii, 66, 136, 181
Rietveld, H.M., 103
 estimation of parameters, 95
Rietveld method, precision and accuracy in, v, 95-107
Riley, D.P., 181
 extrapolation technique, 153, 169
Roberts, J.A., editor, 123
Robertson, B.E., vi, 135, 136
 author, 125-136
Robertson, J.M., editor, 136
Robust/resistant techniques, 135, 183
Rogers, D., 21, 42, 66, 77, 94
 hypercentric distributions, 6, 24, 59, 60, 63
 statistical tests for centrosymmetry, 5
Rohrl, M., 85
Rollett, J.S., 135, 211
 author, 105-107
 discussion of precision and accuracy, 105-107
Rosenbaum, P.R., 211
Rothbauer, R., 181
 quotation of Gauss, 155
Rubidium hydrogen di-*o*-nitrobenzoate, 47, 48
Rubin, D.B., 211
Ryshik, I.M., 42, 66
 series expansions, 38

Sabine, T.M., 103, 195
 atomic scattering factors, 188
 particle-size effect, 103
Sakata, M., 103
Sayre, D., 21, 124
 phase determination, 12
 editor, 135
Scale factor, 3, 92, 110
Scarborough, J.B., 181
 estimation of lattice spacing, 165

Index

Scheid, F., 181
  curve fitting, 168
  min-max method, 156
Schenk, H., editor, 136
Scheringer, C., 211
Sen, S., 181
  data used, 174
Sesquicentric distribution
  orthogonal polynomials and, 90
Sedlacek, B., editor, 22
Shannon, C.E., 112, 123
  information capacity as entropy, 109
  information theory, 113, 114
Shmueli, U., v, vii, 42, 66, 94
  author, 23-42, 53-66
  characteristic functions, 26
  convergence, 29
  moments of structure factors, 24
  probability distributions, 24
  symbolic manipulation programs, 24
Shoemaker, D.P., 195
  influence of individual reflections, 184
Shore, J.E., 112, 123
  argument for minimum-information principle, 115
Sim, G.A., 66
  heavy-atom effect, 43
Sinclair, H.B., 181
  extrapolation technique, 153, 169
Single-Crystal Intensity Project, 183
  analysis of, 187-190
Site occupancies, 205
Skilling, J., 112, 124
  interpretation of probabilities, 110
Smith, A.F.M., 211
Smith, W.B., 211
Solution to the structure problem, existence of, 7-8
Space groups [*not exhaustively indexed*]
  determination of, 79-85
  in intensity statistics, 23-41
Sparks, R.A., 211
Speakman, J.C., editor, 136
Special positions
  effect on intensity distributions of atoms in, 1, 42-51
Srinivasan, R., 50, 66, 94
  heavy-atom effect, 43
Standard deviation
  statistical meaning of term, 107
  validity of computed values, 195
  weighting to reduce, 125-135
Statistical disorder, not necessarily Gaussian, 3
Statistical fluctuations
  distribution of, 87, 88-89
  effect on intensity distributions, 87-93
Steenstrup, S., vi, 123
  author, 113-124
Steepest descents, approximation of distributions by, 87, 93
Stegun, I., 42, 66, 94
  Bessel functions, 29
Stemple, N.R., 195
  tartaric acid, 188
Stewart, J.M., 21, 136
Stezowski, J.J., 135
Stone, C.J., 85
Storey, A.E., 66
Strain determination, 177-180
Strong reflections
  weighting of, 183-195
Structure amplitudes, *see* Structure factors
Structure determination [*not exhaustively indexed*]
  entropy maximization, 109-112
  estimation *via* minimum-information principle, 120-123
  information theory, 113-124
  model for electron density, 115-120
  Rietveld method, 95-107
  weighting schemes in, 125-135

Structure factors
  References to probability distributions will be found under such headings as Intensity statistics, Probability distributions, *etc.*
  distributions of very large, 32-34
Structure problem, existence of solution to, 7-8
Structure refinement, vi; *see under names of methods*
Studden, W.J., translator, 195
Subcentric distribution, orthogonal polynomials and, 90
Substructures, 67-77
Sudarsanan, K., 136
Superstructures, 67-77
  classification of, 67
Symbolic addition procedure in phase determination, 14
Symbolic manipulation programs for moments of structure factor, 24
Symmetry elements
  cause of inaccessible volumes, 44
  effect on intensity distributions, 1
Systematic errors
  effect on intensity distributions, 87-93
  extinction as example of, 87
  in general, 152
  in structure determination, 126 *et seq.*
  testing hypothesis of zero, 151

Tangent formula, 12, 14
  derivation of, 17-19
D(+)Tartaric acid, 183, 187-190
Tauberian theorems, 33
Taxer, K., 77
  algebra of substructures, 67
Taylor, A., 181
  extrapolation technique, 153, 169

Taylor, G.H., 195
  tartaric acid, 188
Temperature factor (Debye-Waller factor), 3, 153
  of β-lactam, 170, 174
  of HgClBr, 170, 174
Thermal parameters
  estimation of, 95, 102, 151
  of β-lactam, 170-174
  of HgClBr, 170-174
Thomsen, J.S., 149
Tiao, G.C., 195
Tikochinsky, Y., 112
Tishby, N.Z., 112
Toutenburg, H., 211
Translational symmetry
  non-crystallographic, 67-77
  See also Bicentric, Hypercentric
'True' intensity, 90
Tsoucaris, G., 21
  conditional determinantal distributions, 19
  inequalities, 15
Tucker, P., 136, 195
Tukey, J.W., 195
  effect of outliers, 185

Urbanowicz [Gałdecka], E., 149

Vand, V., 4
  approximate formula of, 40-41
  multivariate distributions, 37
Varghese, J.N., vi, 123
  author, 113-124
Variance [*not exhaustively indexed*]
  affected by symmetry, 6
  estimation of, 125-135
  in Bond method of lattice-parameter determination, 137-149
  modified, 129
Vincent, M.G., 136
  effect of serial correlation, 128

Wang Hong, vi

Index

author, 125-136
Weak reflections, weighting of, 183-195, especially 191-194
Weaver, W., 112
Webb, G.I., vi, 211
   author, 197-211
   optimization of data collection, 198
Wehrl, A., 123
Weighting schemes
   influence on derived parameters, 183-195
   modification of, vi, 125-135
Weiss, G.H., vii, 66, 94
   accelerated convergence, 26
   author, 23-42, 53-66
   characteristic functions, 26
   convergence, 29
   distribution of large structure factors, 32
   steepest descents, 26
Wells, M., 195
   atomic scattering factors, 188
Welsch, R.E., 195
Whitney, D.R., vii, 181
   likelihood-ratio test, 166
   maximum-likelihood method, 153
Widder, D.V., 42
   Laplace transforms, 33
Wilkins, S.W., vi, 123, 124
   author, 113-124
Willis, B.T.M., 103
Wilson, A.J.C., vii, 21, 42, 50, 66, 77, 93-94, 135, 149, 181, 195
   absolute scale, 3
   author, v-vii, 23-42, 87-94
   bias in $|F|$, 126
   correlation between atomic positions, 44
   distribution of large structure factors, 32
   editor, vii, 66, 136, 181
   errors in angular scale, 138
   errors in scanning mechanism, 138
   hypercentric distributions, 6, 24, 59, 60, 63
   intensity statistics, 1, 5, 21, 43, 54
   moments of structure factors, 24
   non-normal distribution of angular errors, 157
   plot for temperature factor, 153, 156
   statistical tests for centro-symmetry, 5
   vibrational effects, 3
Wilson, K., 50, 195
   correlation between atomic positions, 44
   treatment of weak reflections, 191
Wolberg, J.R., 211
Woolfson, M.M., 21, 42, 66, 77, 94
   approximate formula of, 40-41
   hyperbolic tangent formula, 14, 19
   hypercentric [bicentric] distribution, 6, 24, 59
   multivariate distributions, 37

Yap, F.Y., 149
Young, R.A., 136

Zachariasen, W.H., 21, 195
   extinction correction, 187
   phase determination, 12